THE SEARCH FOR
EVE

Other Books by Michael H. Brown

THE TOXIC CLOUD
LAYING WASTE
MARKED TO DIE

THE SEARCH FOR

EVE

Michael H. Brown

1817

HARPER & ROW, PUBLISHERS, New York

Grand Rapids, Philadelphia, St. Louis, San Francisco
London, Singapore, Sydney, Tokyo, Toronto

Illustrations appear on pages 16, 33, 140, 158, 178, 192, and 209.

FIRST EDITION

Designed by Alma Orenstein

Text illustrations by Diana Coe

Library of Congress Cataloging-in-Publication Data

Brown, Michael Harold, 1952–
    The search for Eve / by Michael H. Brown.—1st ed.
    p. cm.
    Includes index.
    ISBN 0-06-016055-1
    1. Human evolution. 2. Genetic psychology. I. Title.
GN281.4.B76 1990   89-45632
573.2—dc20

---

90 91 92 93 94 CC/RRD 10 9 8 7 6 5 4 3 2 1

To the real Eve,
the one of the firmament,
the one who left no skull

His wonder was to find unawakened Eve,
With tresses discomposed, and glowing cheek,
As through unquiet rest. He, on his side
Leaning half raised, with looks of cordial love
Hung over her enamored, and beheld
Beauty which, whether waking or asleep,
Shot forth peculiar graces; then, with voice
Mild as when Zephyrus on Flora breathes,
Her hand soft touching, whispered thus:—"Awake,
      My fairest, my espoused, my latest
found . . ."

—John Milton, *Paradise Lost,* Book V

And Adam called his wife's name Eve; because she
was the mother of all living.

—Genesis, 3:20

"We don't know what's going on here."

—prominent anthropologist

ACKNOWLEDGMENTS

THIS BOOK was the idea of Craig D. Nelson, senior editor at Harper & Row, whose pen, insights, and friendship always improve and inspire my work. For consultations in person, on the phone, by letter, or all three, for minutes, hours, or days of their time, I thank: Jon Ahlquist, Baruch Arensburg, John Avise, O. Bar-Josef, Lewis Binford, C. Loring Brace, Günter Bräuer, Roy Britten, Wes Brown, Karl Butzer, Rebecca Cann, Steven Carr, L. L. Cavalli-Sforza, Desmond Clark, Hillary Deacon, Arthur Fisher, John Gillespie, Stephen Jay Gould, Fred Grine, Ralph Holloway, Rodney Honeycutt, Satoshi Horai, Clark Howell, W. W. Howells, Arthur Jelinek, Donald Johanson, Richard Klein, Thomas Kocher, Mary Leakey, Meave Leakey, Richard Leakey, Jeffrey Long, G. Lucotte, Alan Mann, Jon Marks, Ernst Mayr, Jeff Mitten, Masatoshi Nei, David Pilbeam, Geoffrey Pope, Yoel Rak, G. Philip Rightmire, Vincent Sarich, Tad Schurr, Henry Schwarcz, Pat Shipman, Charles Sibley, Elwyn Simons, Fred Smith, James Spuhler, Mark Stoneking, Chris Stringer, Randall Susman, Phillip Tobias, Erik Trinkaus, Linda Vigilant, Douglas Wallace, J. S. Wainscoat, Alan Walker, Sherwood Washburn, Tim White, Allan Wilson, Milford Wolpoff, and Ezra Zubrow. My thanks to my family for supporting my efforts while I was away from Manhattan, to my agent, Philippa Brophy, at Sterling Lord Literistic, to Jenna Hull of Harper & Row, and to librarians Joyce Shields, Bill Budge, and Linda

Reinumagi. Also, to "Monkey Business" of Niagara Falls for letting me spend a couple afternoons with a macaque and a chimpanzee; and to Singano, my Tanzanian guide, who drove his hardest on savannah and ridiculous roads trying to help me find Eve and who bore my great demands and endless questions.

1

THEY LIVED where the rhino took dust baths, where the gazelle gathered, and where the giant buffalo roamed. By day they could be glimpsed making their way across the vast savannah, vague upright forms that were decidedly not animals and yet lithely intermingled with them. They moved from tree to acacia tree, foraging among the ambling giraffe, the skittish, dust-kicking wildebeest, the swollen zebras, sticks in hand, wary of lion, their gait purposeful and yet very cautious—a slow and steady and gliding gait. Certainly it was not the stooped and lumbering image commonly accorded to cavemen.

Some carried ostrich eggs and kindling twigs, others water baskets woven with hide and ambary fibers, still others wooden poles or slender stone blades. Their bare feet, padded with thick, leathery calluses, crunched against the tufts of parched grass and haggling brush, toes gnarled by poorly healed bones. Near the foothills a dust devil sucked fine baked clay into its vortex and made its way to the strange pedestrians, who were more advanced than anthropologists would later give them credit for being. A few privileged ones wore crude terracotta-colored cloaks, knotted at the shoulders; others swaths of ragged antelope hide. Others were nude, the dust and ocher their only accoutrement.

Nama, one of them uttered, and two companions looked toward the volcanic highlands. Languidly they scanned a distant jungle that was a Noah's Ark of lizards and birds. Any form of life

seemed possible in the festoons of liana and creepers, in the leopard forest that rimmed the savannah, in the microorganisms that
mutated and multiplied. The sloping terrain, holding runoff from
the heights, was so rich in wildlife—from apes to ants, with buttressed trees, with creeping and flying things—that it fairly whispered the secrets of evolution.

      This was sub-Saharan Africa 200,000 years ago. There were
glaciers slowly moving across other parts of the planet, sheets of
ice that grew like algae from the poles. But Africa was eminently
hospitable, warm and lava-rich; a home, several million years
before, to the first two-legged hominids, the man-apes; then to
ape-men known as *Homo erectus*; now to the very first modern
humans.

      Though there were at least a million very primitive people
living in the Old World, only from this small, isolated population
in Africa did we all descend. They were the only *Homo sapiens
sapiens* at the dawn of modern mankind, and when all their clans
were counted, their number was less than 10,000.

      The rest of Africa and the world beyond it was inhabited
by far more archaic people who bore many resemblances to this
new band in the sub-Sahara but were still too much like the
forerunning *erectus*—low-skulled and beetle-browed. They had
spread throughout Africa and into Asia and Europe, these archaics, many developing into what would be known as Neandertals.
But they were destined for extinction, and it was only the relatively
small band of higher-skulled and gracile Africans, evolving much
more swiftly out of the *erectus* stock, who would serve as our
ancestors. They spread out of the sub-Saharan savannahs and took
control of the rest of the world—wresting it from the archaics—by
30,000 years ago. Just 300 centuries after that, there would be a
nearly inconceivable number of their offspring—5 billion of us—
in every corner of the globe. The last of the ape-men had been
overrun, conquered systematically (and perhaps brutally) by these
fledgling humans from the sub-Sahara. The conquerors had
changed in a unique and fundamental sense. They were able now
to speak and make better weapons, and no longer apelike or
shambling or beetle-browed.

      This was the scenario we were to investigate: that a pocket

of new people from Africa, as yet minuscule in number, would turn out, as scientists now claimed, to be our only direct and common ancestors. It was their destiny, it seems, to reinhabit and reshape the world. From them came the Aborigine and nuclear scientist alike, the Pygmies and towering Watusi, the Egyptians who would take stone-building to new heights on the desert to the north, the Jews up along the sea, the Greeks and Romans, the Samurai and English king, the Russians and Chinese and Americans, the beggar sleeping on a sidewalk grate and the billionaire in a Hyatt hotel.

The original modern humans learned how to escape natural calamities and manipulate the environment. They could build themselves shelters. They could cook. They were organized, their scarce numbers divided into rudimentary clans and subclans, twenty or thirty to a basic, functioning group, living both in temporary camps set in dried riverbeds near the lakes (during this, the dry season) and also in more permanent manyattas out of floodwater reach in the higher land, 200,000 years ago.

It was worth a close look at one such clan, because in it was a woman we have been searching for since the days of Genesis. Alas, according to the scientists, she has been found. Her people still carried remnants of primitive traits, their browridges slightly pronounced, their bones a bit robust. But most had the fully smooth and rounded skull of present-day humans. Some of their noses came to a point, others flared wide and pointy and arching at the same time. They had brownish eye whites and high cheekbones and their skin was dark—sometimes pitch black, in other cases nearly olive. In stature, too, they were like people alive today, but with far more sinew. Even in females, the biceps were prominent, the calves arched with hard muscle, the jaws strong from chewing roots, shoots, and tubers, or a delectable chunk of thick honeycomb.

These ancestors of ours, shrugging off the last remnants of animal behavior, were capable of erecting their living quarters by pounding poles into the earth, anchoring them with piles of stones, and fashioning walls and a roof between them: a thatched, tropical igloo, interwoven boughs plastered with mud and dung, dark and tiny, with room only for a bed or two of gathered grass and the circle of stones to contain a small hearth.

At the center of the huts was a larger structure, nearly forty feet long. It served as the elder's residence and as a warehouse for vegetables gathered by the women. Sapling-walled, it had a wind-screen to protect the hearth and a pit banked with sod to preserve embers. Here the tips of wood spears were hardened by fire. Other tools included hammerstones, scrapers, burins, the elegant new stone blades, and the most rudimentary awls.

Outside, the naked children rolled pebbles into a hole to while away the afternoon as their mothers scraped the fat from fresh hides with sharp little curvatures of rock flaked from lakeside pebbles. There were also heavy choppers and hand axes and grinding stones. Some of the implements were inherited from *erectus*, or even the last man-apes. The stone tools, in their turn, were used to create other implements of wood, horn, ivory, and bone. Smaller pieces of carved wood represented the first glimmerings of ornamentation—a sure sign of self-awareness and individuality. But there was not much time for frivolous labor when survival was constantly at stake. While the men hunted, gone often for two days at a time, the women moved to and from the open brush with hemp-bundled firewood on their backs, or they carved wood bowls and made skin bags before anyone thought they could.

One of these women is the subject of this book.

In a corner of the savannah, six kilometers from the manyatta, the men were cooperating to outfox a gazelle. Several of them encircled it, scaring it into a clutch of other, hidden men who clubbed it and prodded it with their spears. Meat for the hearth. Hides for the capes. Tendons to tie together the smaller pieces of hafted tools and antlers.

Other animals were simply chased until they were too tired to escape, knocked unconscious by a deftly thrown rock, or scared toward a bog where they became mired. When bravery and pride ruled their emotions and the opportunity offered itself, the men sought a lumbering, unsuspecting elephant, ambushing the lone one at the end of a thundering procession and thrusting spears up into its soft underbelly as the other beasts wailed and stampeded off. Larger, sharply pointed hand axes were strung over booby traps to fall upon smaller but equally unsuspecting game. If good-sized meat was unavailable, it was not beneath our ancestors to collect large rodents.

The language was rudimentary vowels. A dozen key ones meant come, go, here, there, food, help, water, danger. Most communication was done with a nod of the head, a sharp glare or glance, a smile or frown. Or by hand gestures. During a break in the hunt one of the men was able to communicate that last year's water hole—a few kilometers to the south—was now fuller than the one they were using this year.

Although the environment was so plentiful that some clans remained in one place, spreading a few kilometers with each new and more populous generation, the progeny of other related clans would find their location marginal and would migrate in search of better hunting or better water holes.

It was during these expansions and movements that they replaced the ape-men.

Some would become nomads, some pioneers. Others would become extinct. The movements, gradual though they were, would reach the point, 5,000 generations hence, where they were up through Israel and into Eurasia: Neandertal territory.

Instead of huts, they often found themselves taking to a convenient cave. The garbage—twigs, ash, old bones, and blunted tools—was removed to another shelter and simply left there. Communication sounded like deep whispers and "ahs" and "oohs" and grunts, but there was also the high-pitched clicking and clacking of children, forming the vaguest of songs.

Their mouths moved up and down and sideways with excitement as they watched older boys, the future hunters, chase a dik-dik into the scrub. Indeed, the older boys could already catch their own hare. They chased the animal until it froze and watched until it bent its ears. When it did, a deft shift sideways had a fifty-fifty chance of being in the hare's panicked route of escape.

Two other teen-age boys were at a rock shelter. They were grinding the tips of slender sticks against a rough patch of granite in such a way as to sharpen them. They were not adept with the weapon yet, but occasionally a small antelope was killed this way and the sticks were the forerunners of metal-tipped spears.

The anthropologist thousands of years later had no way of

knowing this. The wood would never be fossilized. Nor would the clothes or the huts or the modest language last through the ages. Neither would future scholars know for sure that ocher was mixed with fat and rubbed over the body as a shield from the sun and protection against evil spirits.

The older boys were separated from the rest by the elder, who taught them rituals, indicated legends, and tended to medical matters like toothaches, which were solved by simply knocking or chipping out the offending tooth. The elder was squatting beneath a shady, gnarled baobab, eating hard-shelled fruit. He watched as two dust devils rotated along the lower plains, and as he studied the swirling dust he remembered his own hunting youth, how once they had spotted a rare saber-toothed cat. Back in the old days, before more hunters came through, it had also been possible to see eleven-foot ostriches or a short-necked giraffe.

In every modern tribe, the elder was accorded special respect. He exchanged vowels with a woman who passed by with a small parade of children. It was the woman who we have heard about since the First Book of Moses. It really was her. We could hear her talk to the elder, who was actually only forty. She was concerned with the amount and dryness of the wood.

Lambatta—It is still quantity enough, the elder assured her. The woman was highly respected by the men, not only because of her prodigious childbearing but because of her meticulous care of them. And her ability to invent. She was the first to begin the routine of making her children drink herbal juice and eat oranges, and she was the only one who hadn't lost a child to yellow fever and malaria.

She has also devised ornaments of wood shaped like the large hut. It is a way of identifying the clan. They are hung from the neck with a thread of vine.

Dinner tonight is honey and locusts and seeds. The men eat in small groups below the manyatta and the women tend to children near the huts, watching as the sun, just a moment ago so incandescent, begins a precipitous descent such that nightfall is sudden and overwhelming. It is time to think and talk. A fire is rubbed into existence. Grass and twigs are added as the flames leap up, then thick pieces of broken branch are thrown into it. There is no way to saw apart the trunk of a tree but there are plenty of branches to fuel the flames. The rotten ones break like chalk, the twigs send up licks of orange fire, the stones glow so hot you can boil water with them.

Time to think and talk: When will they move the manyatta? Usually every two or three dry seasons, when the easier game, sensing their presence and thinning out, begin to establish themselves elsewhere. It may be time to move then, but increasingly there is the urge to establish themselves permanently.

As darkness deepens a few more men return, sniffing the resinous smoke from fires that glow like coals on the blackened slopes, in the cooling air, under heavens unbridled. There is chatter at the manyatta, the older man staring across the valley to a distant mountain and a startling sight: a strange and previously unseen fire. Someone new has moved in. It is no one they know. And by the next night a scouting mission—one of the older boys who knows the back terrain even in darkness—will be sent to reconnoiter the fire and see if it is a temporary camp and precisely who is inhabiting it.

"Swoopa," notes the elder again. "Malswoopa."

He is of the mind that it is temporary, but he fears violence. Perhaps it is archaics passing through.

Above, the stars are large and clear through the firmament. They still mystify the "old" man. They glow like very distant embers—campfires of the gods. There is chatter around their own fire.

Off to the side, alone, is that incredible woman nursing her young, under a fig tree. One day scientists will look back at her with wonder. Her latest innovation has been to encircle her hut with thorns to keep away nocturnal predators. Especially she hates snakes. They are serpents. *Sanna!* Most other women lose a third of their children before the age of two dry seasons. She is all awareness and instinct. A very special mother indeed. Few women in the history of the world will be as special as she is. Certain of her genes—and her genes alone—will be evident 10,000 generations from now.

Like the other women, she too is callused and hard in the muscle. But she is the most fertile of the women. She has three youngsters at the age of eighteen and another on the way. Two are daughters. Those scientists will one day isolate her genes and nickname her "Eve."

She listens with sudden concern as vervet monkeys spill from the branches behind her, indicating a leopard. The commotion gratefully stops. Now there is only the whistling thorn of dry brush and the distant, staccato rumor of hippos.

It is silly, the name. It will irk a lot of people. But Eve, that's who we're looking for. And in a sense, according to certain of those scientists, this sub-Saharan woman, under the starlight 200 millennia ago, is not just an ancestor or another gracile African— but the world's one common grandmother.

2

THIS, OF COURSE, is a lot of fantasy. It is also the stuff of heated debate. To extensively detail what life was like so long ago, let alone to claim that the scene is the starting point for modern humans, is tantamount, in no few circles, to heresy.

Fundamentalists are not going to like it. Neither are many academics who study such ancient things. They too spit brimstone. They are the ones who sift through old artifacts, compare the dimensions of fossil skulls, and dig in old caves. They are the ones who are supposed to speculate: archeologists, paleontologists, and anthropologists. Their small and select field is known, in the curious lingo that characterizes it, as "paleoanthropology."

But the scenario from Africa comes not so much from fossils, not so much from anthropologists or archeologists, as it does from something even more arcane. It comes from DNA laboratories—the study of those genes that are the blueprints of life itself. Since the 1970s, and especially in the 1980s, geneticists using techniques similar to some of the procedures in recombinant DNA have been tracking a type of inherited material that, they say, came from that single prehistoric woman in Africa. She was old, much older than many paleoanthropologists expected, and she bore only remote resemblances to certain famous and sacrosanct fossils, spawning a lineage that was separate from Neandertals and Peking Man. She was not from anywhere in Asia, which, up to the point of the geneticists' announcement, was still a contender for the seat of modern humanity. Nor was she from Europe, home of Cro-Magnon Man. The one ancestor every living person has in common, they said, and perhaps the only common link in our species, lived those many generations ago south of the Sahara.

Most controversial is the implication that those born of this woman were part of a small and elite band of people who set out to populate the world themselves, vanquishing more primitive and more ape-manlike tribes who many paleoanthropologists have long thought were our immediate forebears. It is enough to make any paleoanthropologist breathe heavily: Finally, by use of some of the more complex techniques known to science, we knew how modern man arose. Finally, a common ancestor of every living being had been located—and, of all places, in a test tube! Finally, we knew what paleoanthropology had not been able to tell us and what many in the field did not expect: that the human tree for modern men does not stretch into the unforeseeable past, but could be traced back to a specific era.

Finally we did not have to rely solely upon fragmentary fossils, but instead upon living genes.

No one had made quite such a claim before. A single woman whose genes had survived through the centuries. It was a eureka moment: she lived about 200,000 years ago, probably on the savannahs of South or East Africa, as in the previous chapter. And yes, her offspring had conquered the world, developing technologies that tamed nature and allowed them, and them alone, to plant themselves permanently across a previously forbidding, ape-man planet. It was the type of discovery paleoanthropologists could not have settled upon alone, and though it received far less publicity than many fossil finds—at least at first—the vision it set forth was as riveting as anything in a century of cave digging: A lone woman, baby in arms or strapped to her back with crude cloth, strong-muscled and dust-covered, perhaps (as one publication, *Newsweek*, suggested) as muscular as Martina Navratilova, aware of the wild dogs and the snakes and the leopards in anything but a Garden of Eden.

This was our Eve. It was also the newest and hottest controversy in paleoanthropology.

The tumultuous world of paleoanthropologists was

thrown, worldwide, into that much more turmoil. For decades the business of deciding where mankind originated—at least at the scientific level—had been securely in the hands of paleoanthropologists, who drew Y-shaped trees to demonstrate the progression of modern humans from the apes, from the bipedal hominids (like man-apes and later *erectus*), and from the Neandertals (which until recently had been spelled with an "h" after the "t"). It was not the terrain of geneticists in white lab coats. And what those geneticists were stepping upon was a minefield of raucous history.

The tradition of archeology and human fossil evidence could be traced at least back to the seventeenth century, when Isaac de la Peyrère, from France, began studying an odd assortment of stones that looked as if they had been chipped by hand long ago. Their unusual shape implied not the forces of nature—not wind, earth pressure, heat, or rain—but instead the fashioning of stone into tools by primitive men who, Peyrère postulated, lived in the time before Adam. In 1655 his findings and theory were subjected to great disapproval, including, of course, the standard ritual of the day: a public book-burning.

Sixteen years later, even stranger things started turning up in Germany. Human bones were discovered alongside those of extinct bears. That hinted at just how old those bones were. In fact extinct animals began showing up throughout western Europe. In 1796 Georges Cuvier, another Frenchman, found the remains of ancient mammoths and gigantic reptiles, and pretty soon paleontology (the specific study of fossil animals, though often used in the same way as "paleoanthropology") was coming into bloom, with techniques to unravel the secrets of the long-gone past by counting and categorizing geological strata.

But those human bones associated with extinct animals: were they mystical beings from the underworld, or long-gone Druids, or Antediluvians? Perhaps the bones and fossil shells and oddly chipped stones—the fashioned flint that was found in beds of gravel and below stalagmite floors—were all just a supernatural ruse, planted for the sole purpose of befuddling mankind.

Then, in 1856, three years before Darwin propounded his theory of evolution, there was the absolutely momentous discovery, in Germany's Neander Valley (near Düsseldorf), of a chert-

littered cave deposit—Feldhofer Grotto—that contained the remains of what was soon recognized as a fossilized human skeleton.

It was the discovery of Neandertal Man and the beginning
of real paleoanthropology.

Upon closer inspection, Carl Fuhlrott, a local mathematics
teacher and amateur natural historian, had realized that the skeletal remains possessed a number of peculiar traits that defined them
as very ancient. The eyebrow ridges were heavy and slanted, there
was a deep depression at the root of the nose, and there was a low,
narrow, sloping forehead, which Hermann Schaaffhausen at the
University of Bonn assigned to a "barbarous aboriginal race."

Or were the strange, stout bones a result, as was also suggested, of childhood rickets followed by arthritis in old age?

If he was bowlegged, that might have been from riding a
horse, still others thought. As for the unusual browridges, it was
guessed that the pain from rickets had caused the savage man to
pucker his brow in a permanent frown that eventually ossified.

The debate was on and continues to this day. At issue was
and is where or whether such archaic humans fit in our lineage.
Originally, however, the argument also concerned whether or not
the skullcap that was found near Düsseldorf belonged to a race
that was extinguished during one in a series of cataclysms that
punctuated history (as Cuvier believed), or whether, as the unifor-
mitarians proposed, it was the result of a world which, since creation, had been subjected to gradual modification by the action of
natural agencies. In other words, it provoked the seminal debate
between evolution and supernaturalism, followed by a century of
subsequent debate about whether this being from the Neander
Valley—this Neandertal, as he is now known—was ancestral to
modern humans or merely a freakish and now extinct offshoot: an
odd and aimless type of ancient European.

The issue has never been resolved. Men like Marcellin
Boule, a leading French anthropologist just after the turn of the
century, conjured an image that would live through to the present
day, of a stooped, shaggy-haired, and bent-kneed brute with a
retreating forehead and bulge in the back of the skull known as
an "occipital bun"—along with those obvious browridges. It was
the classic and unfair image of a club-wielding caveman—uncouth

and repellent, in quest of fire, dragging a female by the hair and back to the cave. Boule considered the Neandertal "an inferior type closer to the Apes than to any other human group." No Adam or Eve here. Fuhlrott himself was inclined at first to venture that the remains were those of an unfortunate mortal washed into the grotto during Noah's Flood. There were also allusions by scientists like Schaaffhausen to the chance that the fossils derived from the legendary wild races of Europe, which had been mentioned as far back as Virgil in the *Aeneid*.

Such a beastly visage did serve, however, to reinforce the notions of evolution and natural selection, which were soon all the rage. Darwin's great defender, Thomas H. Huxley, took a close look at the Neander Valley skull and concluded that while it had many features of a Patagonian or even a modern European (especially as regarded the size of limbs and brain), it was the most apelike human cranium yet discovered. At the same time, however, Huxley did not consider it an intermediate between man and gorilla. Instead, he viewed it as a human specimen that had reverted back to a few gorilloid traits. This was man, not monkey. It wasn't even the primeval man everyone was soon looking for. Neandertal Man would one day be designated as a subspecies of *Homo sapiens*—a far cry from a different genus. Its current classification is *Homo sapiens neanderthalensis*. Whether related to Neandertals directly or not, we are another subspecies, as *Homo sapiens sapiens*, of the same general family.

"In still older strata do the fossilized bones of an Ape more anthropoid, or a Man more pithecoid, than any yet known await the researches of some unborn paleontologist?" Huxley asked prophetically.

For a Dutchman named Eugène Dubois, who was born two years after the Neandertal discovery, this fossil was an inspiration to search for a more primitive missing link that would build the true primeval bridge between man and ape, just as Huxley suggested. Dubois, a doctor and instructor at the Royal Normal School in Amsterdam, specializing in anatomy, envisioned this hypothetical skull as lying in limestone deposits or a tropical cave. His reasoning was that prehistoric man would not have fared so well in Europe during the glaciers. One of his teachers, the zoologist Ernst Haeckel, had predicted that the sought-after link would be found either in Africa or "Lemuria"—better known today as southern Asia. A warm, forest-clad terrain, moreover, was where Darwin had predicted man's ancestors would be.

Haeckel also had imagined that the manlike ape would have hair covering the whole body, arms longer and stronger than ours, legs that were "knock-kneed, shorter and thinner, with entirely under-developed calves; their walk but half-erect." Neandertal Man may have been crude and a bit hunched, but what was envisioned now—and what Dubois ventured, at the age of twenty-nine, to find—was nothing less than half-man, half-beast. He set out for Sumatra and then Java toward the end of the nineteenth century as a health officer in the Dutch colonial army.

Incredibly, Dubois, in many regards, found just what he had set out to find. By 1890 he had discovered a piece of ancient jaw, but that paled alongside his discovery the following year, near the village of Trinil, of a skullcap that he was sure was the missing link. In 1892 he unearthed a human femur to go along with it. While the thighbone looked much like those found in modern humans, the skull—smaller-brained than a Neandertal but much larger than an ape's—was the in-between Huxley had predicted. If the Neander specimen had very prominent brows, those of the new discovery in Java were absolutely massive. Similar specimens found later on would show the ridging around the eyes sweeping from brow to ears like a pair of thick, bony, built-in glasses.

The more Dubois studied the Java fragments, the clearer it became "that they are really parts of a form intermediate between man and apes, which was the ancestral stock from which man was derived," he wrote in 1896. Dubois called *his* skullcap, which everyone agreed was more primitive and certainly more apelike than the Neandertal, *Pithecanthropus erectus,* meaning an erect-walking ape-man.

"Apelike man or diseased man, the native of the Neandertal has from the very first always been considered as an undoubted, real man," wrote Dubois. "The human character of the *Pithecanthropus* is, however, very questionable. The skull of the gibbon almost doubled in size would not be very different from it in appearance."

However much the unrefined skull seemed, at first, to an earnest Dubois, as resembling that of a huge ape, the distinguishing features remained its brain size and the important fact that, unlike the chimpanzee or gorilla which Huxley had said were our closest living relatives, this creature from Indonesia had strutted about on two feet. When, half a century later, it was realized that, as was the case with Neandertal, the apelike features of "Java Man" had been overstated, *Pithecanthropus erectus* was placed, like the Neandertal, into the same genus as current mankind. This time it was classified, however, as a separate species. Java Man was human, *Homo*. And *Homo erectus* was the new scientific name.

The skull is now believed to be 800,000 years old, which is substantially older than the 30,000 to 125,000 years frequently assigned to Neandertals. And it meant, obviously, that primitive humans had been way over in Asia before Europe. Naturally Dubois soon found himself in the midst of a wrenching controversy. So embroiling was the debate that he was to hide the fossils under the floorboards of his dining room for something like three decades. He and Java Man, the now-famous *erectus*, predecessor of even the Neandertal, the closest anyone had yet come to the missing link, simply disappeared from public view. The vitriol started in the form of doubts that all the bones Dubois had found belonged to the same creature. Couldn't ape remains have been mixed in with those of a human?

This creature simply didn't fit the expectations of science, which had predicted that the missing link would have a humanlike head but the body of an ape. Java Man was just the opposite: the apish traits were in the skull, while he walked, like men, on two legs.

Englishmen tended to think of Java Man as a human with apelike features. Germans saw him as an ape with the attributes of a human. Ministers denounced the brute from the pulpit while assuring their congregations that Adam and not Dubois's weird skull remained mankind's first ancestor.

But *erectus* went on to great respectability, and it has remained for a century now as a widely accepted ancestor of the human line. In time it would be just about unanimously accepted

A standard illustration on museum walls and in high school textbooks, these reconstructions suggest far more concrete knowledge of human evolution than we actually possess.

as a rough draft or prototype for both Neandertals and modern humans.

The predominance of *erectus* in our lineage was vastly reinforced from 1927 to 1937 as a treasure chest of similar *erectus* fossils was dug and blasted out of limestone caves southwest of Beijing at a place known as Zhoukoudian. Parts of more than forty people were found, along with thousands of stone tools and evidence that *erectus* used fire and may also have been a cannibal. "Peking Man," as the fossils became known, was very much like the Java *erectus*. But he had a slightly larger brain and a slightly heavier jaw. It seemed to constitute incontrovertible proof of mankind's lengthy existence. Although these famous fossils, which caused every bit the stir Dubois's fossils had a quarter century before, were lost during the rush to get scientists out of China as Japanese soldiers surrounded Beijing (or Peking) during World War II, splendid casts of them were made and have been dated at 220,000 to 500,000 years—a slightly modern version of the Java *erectus*.

So, after the turn of the century, and especially after the fossils from China confirmed *erectus*'s humanness, we had a horizontal chart, in encyclopedias and on classroom blackboards around the world, that showed the progression from ape to Java and Peking Man to Neandertals and modern humans. By that time, however, another discovery had been made that would have caused Dubois and Haeckel to salivate. It was a find that would revolutionize concepts of how and when hominids got up on two feet. In South Africa an anatomy professor named Raymond Dart found not another ape-man but something older: a man-ape.

Neandertals and *erectus*, as I've stressed, turned out to be much, much more human than apish. That is why I choose to use the word "ape" as a mere adjective for what they really were: men. There are variations among current humans that are nearly as great as some of the variations between *erectus* and certain modern people. The real "missing link" (though of course there *is* no real missing link) seemed to be a young creature discovered by Dart from a limestone quarry a couple hundred miles from Johannesburg. The find came in November of 1924, the same year as the

Scopes Trial, and the creature is known, because of its inferred age (three to seven years old), as "the Taung child."

Taung was not a him or a her but more of an "it"—an ape with some characteristics of a human. And it was followed by the discovery of similar man-apes in Africa. If Neandertals had thick bones and a projecting, coarse face, if *erectus* had an even coarser face and a flatter, lower vault, these were indeed minor variations in comparison with the Taung child, which, except that it was much smaller in stature (less than four feet tall), resembled the creatures portrayed in *Planet of the Apes* more than it did a human.

Where *erectus* and Neandertals were humans with some flashbacks to the apes, Taung, it should be stressed, was the reverse: an *ape* with certain distinct human characteristics. It had a small brain—half the size of *erectus,* whose brainier specimens were similar in size to many modern humans—and the long, thick jaw of a female gorilla (though somewhat shorter and lighter than a gorilla's). *Erectus,* Neandertal, and current people were *Homo,* human. Their brain sizes often overlapped. Taung was clearly more ape, with a brain only slightly bigger than a chimpanzee's. The nose was flat. The jaw dominated the face. There was a gaping, thrusting mouth.

Whatever anyone wanted to call the Taung child, it belonged in a zoo more than it did in a day-care center. But Taung was certainly not *totally* apelike. Instead it seemed to represent, in many regards, a true intermediate or transitional creature between the gorilla and *Homo.* Taung had humanlike teeth instead of a gorilla's long, fanglike canines. It also had a bit of a forehead. Most importantly, it didn't get about on all fours. It was a primate that walked upright. Its spinal cord entered the brain not at the back of the head, like a gorilla's, but at the bottom of the skull, suggesting bipedalism, the two-footed gait. Although that didn't make it human, it allowed it to slip into the vaguer category of "hominid."

For sixty-five years the Taung child, like Neandertals, like Java, has remained a matter of emotional controversy. At first its existence was hardly acknowledged. Dart and his skull met stony resistance from stuffy and jealous colleagues, and when the Taung specimen *was* mentioned, it was often in the form of condescending cartoons and music-hall jokes. The Roaring Twenties roared in laughter at this new contender for the "missing link" title. People still argue about the same things they did when the discovery was first announced. There are those who believe it is little more than an ape. And while Taung has enjoyed periods during which it was automatically catalogued into mankind's lineage, there is now a growing movement to again shunt it aside as an extinct hominid that was not one of man's direct ancestors.

But suddenly, in Taung, there was an ape that walked on two feet a million or so years before Java men were around—a creature far more chimplike in appearance, the closest we had come, up to that point, to bridging the yawning gap between mankind (even primitive mankind like *erectus*) and the apes. This creature was most definitely no Eve. For at the same time it was considered a hominid (as opposed to a pongid, the family of apes), it was nonetheless assigned to an entirely different genus than man, one known as *Australopithecus*, meaning, contradictorily enough, "southern ape." No *Homo*, perhaps, but a discovery that ranked in importance with both *erectus* and Neandertals.

And it started eyes turning to Africa.

Before *Australopithecus* came along, southern Asia, especially the ethereal heights of the Himalayas, had been considered the most logical candidate for the origin of man. It was Asia, after all, where the hulky cavemen were being found, the *erectus* that were hundreds of thousands of years older than Neandertals. It was Asia that had the vegetal variation that seemed so seminal and primary to life; the required warm climate; and apes like the gibbon and orangutan.

Most importantly, it had location. Asia was absolutely centrally fixed. It was contiguous to most other landmasses, reaching a finger even to touch North America near the Arctic Circle. The very name that the Chinese called their country—*Chung Kuo*—meant "Middle Country."

It had been the seat of some of the oldest civilizations, where mankind had first developed gunpowder, porcelain, printing, and paper. In China, the continent of Asia also possessed the single largest population of *Homo sapiens*, and as a whole, Asia claimed more than half the world's people. It was certainly a logical place to look for an Eden.

But Taung disrupted the idea that Asia was our birthplace and drew fossil hunters to the Dark Continent. We were no longer talking about an ape-man that lived hundreds of thousands of years ago but of one that went back a million or so years and truly *did* have the features of an ape. The shift of attention to Africa as the possible cradle of civilization was enormously bolstered by the discoveries of the Leakey family. These fossil hunters from Kenya—mother, father, and son—were to find other australopithecines and older *Homo*s (such as *Homo habilis*) and even an *erectus* that was dated earlier than the specimen from Java.

In all these fossil finds, however, in the seeming progression from the apelike *Australopithecus* to the less apelike *habilis* to the manlike *erectus* and Neandertal Man, there was virtually no indisputable evidence of precisely where and when *Homo sapiens sapiens*—anatomically modern humankind, whose repetitious *"sapiens sapiens"* distinguishes them from old *"archaic sapiens"*—first arose. The fossil record was obscure and at times outright blank. It was fairly certain, in the minds of evolutionists, that *Australopithecus* had turned into *erectus,* and that *erectus,* after originating in Africa, had then spread around the Old World. But it was not the origin of *erectus* we were shooting for. *Erectus* was just the rootstock, ancestral both to us and Neandertal Man. We wanted to know exactly where *erectus* had turned into *Homo sapiens sapiens.* Somewhere in the hordes of ape-men there had to have been a clan that was first to turn modern.

Was modern mankind only 6,000 years old, as the religious fundamentalists believed, or a million years old, as old-time anthropologists preferred? Did our particular subspecies originate in Europe, where the modern-looking, cave-painting Cro-Magnon Man began around 30,000 B.C., or perhaps even in North America or Australia, as some dared to speculate?

Or did round-skulled, big-brained, small-browed humans—the very first anatomically modern humans, like the ones characterized in the first chapter, with no pronounced apelike features—evolve in several regions independent of each other?

Was *erectus* as best personified by Java and Peking Man the true founding stock, as so many scholars had come to accept? Was Neandertal an ancestor of modern humans or an extinct offshoot?

Again, where and how did the first *modern* humans arise to take over the globe, causing the less human hominids to head for the dustbin of extinction?

Enter the geneticists and the dramatic claims of the late 1980s. Into the fray of paleoanthropology, onto that emotional and venomous intellectual stage, against the backdrop of ape-men, came the lab coats and centrifuges and nucleotides. Of all intrusions into the clubby and rancorous world of Indiana Jones: geneticists! They were now going to discover what the fossil hunters couldn't, what even the Leakeys could not find: the very first point of true modern origin, and they would also specify the time.

3

It was on new year's day that the geneticists staked their claim. On page 31 in the January 1, 1987, issue of *Nature*, two scientists from the University of California at Berkeley and another from the University of Hawaii presented a highly technical paper entitled "Mitochondrial DNA and human evolution." Under the heading was a summary (known as the "abstract") that made all of paleoanthropology take note. It was even startling: "Mitochondrial DNAs from 147 people, drawn from five geographic populations, have been analyzed by restriction mapping. All these mitochondrial DNAs stem from one woman who is postulated to have lived about 200,000 years ago. . . ."

While most of the rest of the world was unfamiliar with the article (and was recovering from the night before), the diverse worlds of anthropology and genetics were immediately seized by the significance of the piece. *Nature* was not quick to accept new claims, and the journal had the first and final say in many matters of science. Though the average person had never seen the magazine (with an international circulation of just 40,000, it wasn't to be found at the corner store), *Nature* was a bible of new biological, astronomical, chemical, and anthropological assertions—the most prestigious science publication in all the world. It was where the first identification of the AIDS virus was announced, where the discovery of lasers was originally explained, and where James Watson and Francis Crick had first propounded the very structure of

DNA itself. One could go back decades, at least to the announce-ment of the Taung child, and note its lofty paleoanthropological role. It was where debate over the status of man-apes—the aus-tralopithecines—still took place. It was where the Leakeys an-nounced many of their great fossils. Intense commotion could be caused by technical reports that were hardly more than a page in length. Crucial intellectual debates were spawned by mere letters to the editor. Scientists chewed every morsel like cud.

This report ran for five full pages, tucked between a cover story on light emission from the hydroid *Obelia* and a report on optical computers. That was fairly good play. But, as with many of its articles, *Nature* had waited a good while before publishing it. Almost ten months of agonizing debate went on, ten months of struggle to reach a consensus on whether or not the article, with its loud and hypothetical claims, should be published by such an august journal. It was sure to generate some newspaper articles— location of a woman who inevitably would be referred to as "Eve"—and though it was hardly able to compete with the top stories of the week (the Iran-Contra affair was coming into bloom, and in the science world all the talk was of superconductivity), the report, crammed with lab results and the spectacular claim of finally locating man's one common ancestor, was nonetheless des-tined to spark years of argument. Such was always assured in the small, aloof, and extraordinarily contentious world of paleoan-thropology.

If the average *Homo sapien* was oblivious to the report (it didn't exactly make a news bulletin during the football games), that didn't detract from the fact that science was suddenly taking a wholly new approach to the history of its own species. What the article explained was that, in analyzing DNA from the energy-producing compartments of the human cell known as the mito-chondria, the authors, Rebecca L. Cann, Mark Stoneking, and Allan C. Wilson, had discovered that this genetic material, inher-ited through the ages, provided "new perspectives on how, where, and when the human gene pool arose and grew."

From samples of placenta, they had extracted and purified enough hereditary material to compare the intricate variations between and among peoples of Asian, European, African, Australian, and New Guinean descent. The mitochondrial DNA—or "mtDNA," as they insisted upon calling it—offered a wholly unique attribute that couldn't be found elsewhere: it was inherited only maternally, from the mother. Except for occasional minor mutations, this type of DNA was passed intact from great-grand-mother to grandmother to daughter with virtually no input from males and thus no mixing—no blending of father's and mother's genes—that would jumble, complicate, and thus obscure its history. It survived the generations—the millennia—without being fudged by recombination. And for that reason, explained the paper, mitochondrial DNA—as opposed to nuclear DNA, where such blending does occur—was a powerful new tool "for relating individuals to one another."

In sum, mitochondrial DNA was a convenient tool because it was much smaller and simpler than the DNA in a cell's nucleus. While the nuclear DNA, composed of perhaps 100,000 genes, had not yet been mapped to spell out the entire message hidden in its chemical codes (and is currently the subject of a $3 billion federal effort, expected to take fifteen years, aimed at just that), the DNA in a cell's mitochondria had already been fully mapped and found to contain but thirty-seven genes, all of them inherited solely from our mothers (instead of the complicating mix from both parents that, again, occurs in the nucleus). "So you're connected by an unbroken chain of mothers—whether you're a son or a daughter—back into the past," explains one of the authors, Allan Wilson. "And what we're interested in doing is building up a genealogical tree that connects those maternal chains."

Although the metaphor is anything but perfect, for our uncomplicated purposes we can visualize DNA—or deoxyribonucleic acid—as the biological equivalent of computer punchtape. The "holes" in the punchtape are codes which direct the cell to form the proteins that serve as the foundation for our bodies. Certain portions and sequences of the computer tape are what we call the genes. Besides directing the daily activities that keep the body humming, DNA determines our heredity: height, skin pigmentation, and eye color down the generations, the transmissions of traits that keep the race or species going for thousands or millions of years. DNA is, in fact, literally a blueprint for life itself.

Where DNA in the nucleus, as Cann said, may determine everything from whether we have curly hair to whether we can curl our tongues, the role of DNA in the *mitochondria* is much more limited but nonetheless extremely vital: it codes for key components of our energy-production system. Put another way, it is a piece of coded information in the mitochondria, and the mitochondria are bean-shaped structures that, in their turn, are the power plants of the cell. It is the task of these microscopic components to extract energy from food molecules floating in the sappy cytoplasm outside the nucleus.

While it is the nuclear DNA that determines what the next generation will look and perhaps even act like, the DNA in the mitochondria—circular-shaped and densely packed with its own brand of genes—has several characteristics that make it more valuable in studying evolution. In addition to the fact that it's simpler and maternally inherited, allowing a geneticist to look straight into the female past, it grants a magnified view of genetic diversity because mutations—changes in the tiny biochemical configurations that happen as a matter of course—accumulate much faster in mitochondrial DNA than in the nucleus. They give each individual a "signature" that can be compared with others. (Most evolutionary change, it is believed, starts with mutations in genetic molecules.)

But many mutations have little or no effect on the functioning of our organisms. They are "neutral." They seem to occur randomly. And they accumulate over time, changing ever so slightly the configuration of certain of those "holes" in the "punchtape." By zeroing in on the patterns, geneticists, after relating one person's DNA to another's, can supposedly determine how close people are, and can then represent them as twigs on a genealogical tree. "If you're good enough a molecular biologist," says Wilson, who clearly thought he was, "you can reconstruct that branching diagram leading back to that one mother by taking a count of the number of mutational differences among the twigs. We build genealogical trees by comparing mitochondrial DNA from the terminal twigs connecting through unbroken chains of mothers back to the mother of us all."

The geneticists felt, in effect, that they had come upon a

means of studying the origins of modern man without resorting to chips of flint and scraps of fossil. They could even date backward (far backward, without studying paleontology or geology), for the mutational differences (or "divergences") between people that occurred, they calculated, at a set rate of 2 to 4 percent every million years. In other words, small but regular changes in the DNA through the ages functioned as sort of a molecular clock. The ticks were mutations, the minutes were many centuries.

It seemed, in mitochondrial DNA, that a new window— really a looking glass—had opened onto previously impenetrable history.

As Wilson goes on to explain, the key findings were "that these mutations, the neutral ones, accumulate at much the same rate in all organisms ranging from bacteria to plants to animals, that the same basic rate of ticking, the same rate per year, is observed in all of these creatures, so in this vast array of organisms, we have a timepiece ticking away in a statistical fashion (not like a metronome), and allowing us the possibility of being able to put a time scale onto all of evolution, regardless of whether the species have a fossil record."

So powerful was the new technique that it was "entirely possible," said another of the geneticists, Rebecca Cann, "that we could identify Cleopatra's mitochondrial genes in modern people, and, at least theoretically, also specify her entire mitochondrial genotype."

More to the point, they could look for the genes of cavemen in a future Yuppie.

And whatever population had accumulated the most mutations—was the most divergent among each other—could be assumed to have been the oldest. The longer a population had been around, the more mutations.

Such a possibility was enough to cause paleoanthropological angina. It was one thing to be hit with a process that seemed frighteningly complex, but then for these geneticists to take a new, nearly unknown technique—a technique the vast majority of anthropologists were totally unfamiliar with—and swiftly pronounce a major evolutionary find based upon that technique, well . . . There is nothing more controversial to a paleoanthropologist than a claim that someone can precisely describe when and where and how man arose from a primordial, ape-filled past.

As frequently as not, paleoanthropologists can't even agree on the most basic things. Often they see different traits, different species, and far different implications—sometimes even a different genus—in the very same cranium.

Each major anthropologist has a highly opinionated way of interpreting fossils and a highly individualistic way of presenting evolutionary trees. When two or three of them agree on an issue (or *part* of an issue), this becomes a theory or "school" of thought. That's all it takes. And the ground is constantly shifting. "You can't really tell who thinks what this week or last month," comments one such eminent paleoanthropologist, Clark Howell. "The consensus is who shouts the loudest."

In short, it's bad enough when paleoanthropologists squabble among themselves, but now outsiders—geneticists—were in effect proclaiming—shouting—that they could solve what fossil hunters and anatomists had never been able to resolve themselves: the period of time during which primitive or "archaic" humans transformed into fully modern beings.

With equal audacity, the geneticists were going to say *where* modern man arose, with an air of unprecedented authority. It is difficult to describe the emotions such an assertion can provoke. While paleoanthropologists had pretty much reached a consensus that Africa was the place of birth for the oldest of the man-apes and ape-men, the place of origin for the next major stage in human development, anatomically modern men, was still very much up in the air. Since the old caveman *erectus* and his nearly equally beetle-browed descendants had spread pretty much around the Old World—Africa, China, Indonesia, Europe, Israel—the transition into modern human form could have occurred in any of those places. From the raw ingredients of *Homo erectus* (which, miraculously enough, nearly everyone seemed to agree was the major precursor of the sapient strain), evolution appeared to have reached sort of a critical mass, erupting here and there first from *erectus* into the slightly less primitive "archaic" *Homo sapiens* (which included Neandertals) and then into the nearly browless, smooth-headed, small-faced *Homo sapiens sapiens.*

All Eurasia and Africa were therefore the stage. Though best symbolized by Eugène Dubois's finds in Java, *Homo erectus* or archaic descendants were now even known to have existed in India and the Soviet Union. Up until 400,000 years or so ago, *erectus* was the dominant caveman—hunter, maker of fire, uttering perhaps a few rudimentary but decipherable sounds. He was the caveman's caveman, granddad of Neandertals and us as well! And the very "type" specimens for *erectus*—the fossils looked upon as those which defined the species—were the ones composing Peking Man and Dubois's monumental finds in Java.

So for many years an overshadowing dispute had been the basic one of modern origins, and debate was intensifying. There were those who believed anatomically modern man evolved over a broad geographical front. That is, that Chinese *erectus* gave birth to modern Chinese, that African *erectus* fathered modern Africans, and that the old cavemen of Europe led to modern Europeans. They were all connected to each other by occasional sexual intermingling.

An *erectus* from China, in other words, might have a fling with an *erectus* carrying African genes.

This was the "multi-regional" model.

And there was no single Garden of Eden.

The opposing school of thought conceded readily that *erectus* was our precursor, but did not accept that a wide range of these primitive men evolved *simultaneously* into fully modern beings. Instead, adherents to this school believed a select population of the *erectus* assumed modern characteristics in one region, and then simply replaced the more primitive types in other regions. This school of thought is known, because of the idea that a small group served as the founding population, as "Garden of Eden." Although the nickname is an obvious one, it manages nicely to convey the idea that we rose from a single, rather confined, even isolated population of *erectus* that then fanned out from its original locale. Those who adhere to this proposition tend to believe Neandertals were an extinct and meaningless side branch of humanity, disappearing with nary a trace, overflooded by Eve's offspring.

No one had been able to settle the issue once and for all. The biggest problem for the Garden of Eden school, proposing as it did a single point of origin, was proving where that single point was. There was that void in the fossil record. Or, as always, the bones told conflicting tales. While Africa had been a contender for the seat of humanity ever since the Taung finding, that continent was considered by many to figure most prominently only in the far past—and it was a long way from bipedal apes to Java Man. "The idea of an African origin for anatomically modern man was still not quite there," says Howell, an especially influential paleoanthropologist who teaches at Berkeley. "There was the conceivable possibility, but the evidence was still so scrappy, the dating was so equivocal or unconvincing or inadequate, that people were probably unwilling to hang their hats on that stance. Eurasia was still a consideration. India was a possibility talked about."

The candidates even included Arabia.

Europe, where cave art hinted at the transition to modern man in the form of Cro-Magnon, and Asia, where *erectus* had been found to be at the very least half a million years old, were, meanwhile, still in the running as places where *erectus* had evolved into precious *Homo sapiens sapiens*. Because no one seemed to doubt the pivotal role *erectus* played in the human lineage, and because its most famous representations were in Asia—once again, the Java *erectus* and also the *erectus* known as Peking Man—there was a strong tendency (despite growing leanings toward Africa) to keep Asia as a bona fide candidate for the cradle of modern mankind. There seemed to be a continuity in Asia from half a million years ago to the present populations. Even the Garden-of-Edeners conceded some gene mixture.

The Aborigines of Australia, who must have gotten down there from Asia, actually *looked,* to some scholars, like conceptions of old Asian *erectus*. Thick brows. Strong at the jaw. Africa was fine as the rootstock of man-apes and ape-men, but the delicate evolution to *Homo sapiens*—millions of years after the first australopithecines had been spawned—seemed better placed in areas where civilization was most advanced.

Europe was also still in the running, favored by a few scholars who resided in England, Germany, and France, but the real debate was Asia versus Africa. The most famous and best-loved fossils of *erectus*—the very skulls that had defined the *erectus* species, for many years the most ancient human fossils anywhere, or at least the ones known for the longest time to the general public as most ancient—came not from Africa but Asia. Franz Weidenreich, a German who was perhaps the most meticulous describer of fossils ever known, believed that development into modern form was the destiny of a number of major lineages over that broad geographical front I mentioned. Those who followed Weidenreich figured each region had its own distinct rootstock which had evolved to form the various races and ethnic groups that exist today, and though there was some mixture and migration between the rootstocks, remnants of the original, primitive people could be spotted in modern populations. Besides the European genes derived from Neandertal, Weidenreich's disciples, especially the American anthropologist Carleton Coon, envisioned Peking Man leading to modern Mongoloids (an idea enshrined in the *World Book* encyclopedia), while Java Man led to those Australian Aborigines. The Aborigines, thought Coon, were so thick of brow it looked like they were still sloughing off *erectus* traits. Coon went so far as to conceive that evolution into modern man developed in each major part of the world nearly independently.

While no one could prove that Asia was our first homeland, neither could anyone *disprove* it. "Paucity of fossils and infirmity of dates," repeated another noted authority of the day, William W. Howells, "remains a central problem." No one could even discard Europe once and for all—no one until now. Though, in true scientific form, they were going to be ponderous enough to appear a little bit conservative, the geneticists, writing this New Year's Day in *Nature*, did not mince too many words in venturing their key claim: that modern man arose, alas, in sub-Saharan Africa.

Forget the nonsense about Aborigines looking like primitive Indonesians. Australia was colonized by Africans. *China* was conquered by Africans. Europe had been taken over by Africans. The Neandertals? They weren't direct ancestors. Their genes had not even been incorporated into those of the advancing Africans. They weren't the precursors of modern Europeans or anyone else, despite what many paleoanthropologists continued to believe. They had been swept into complete extinction, warranting little more than a footnote, irrelevant in the grand scheme of things.

Instead, every existing human could be traced back to a very special colony of humans from the deep, dark continent of Africa, said the geneticists. They had found, it seemed, the oldest maternal lineage leading to modern-day humans! Everyone had recent African roots, whether the leather-faced Indians who first crossed the Bering Strait or the milk-white Swedes.

The geneticists set the age of "Eve" (their predictable nickname for the one African woman whose mitochondrial DNA they had tracked all the way down the ages) at between 140,000 and 290,000 years. They rounded her median age to 200,000. She was almost surely a highly evolved descendant of an African *erectus*—as opposed to the more famous Asian variety of *erectus* like Peking Man. And her offspring, in turn, helped found a population that was growing increasingly sapient and increasingly migratory, sloughing off old *erectus* traits and spreading out of Africa perhaps 90,000 to 180,000 years ago to take over the world, although it was warned that the migration may have been as "recent" as 21,000 B.C. In other words, a population of *erectus* in a single and isolated spot had made the momentous change into modern humans, displacing the other citizens of our planet like a fast and colorful and aggressive fish—a beta—assuming complete control of an aquarium.

There was little evidence, said the geneticists, that this advanced African population interbred with existing and more primitive populations in Asia and Europe. Instead, they simply outbred and conquered them. They must have possessed some vast superiority, and perhaps a streak of real nastiness. "Thus we propose," said Cann, Stoneking, and Wilson, "that Homo erectus in Asia was replaced without much mixing with the invading Homo sapiens from Africa."

Invading! The very word sent the imagination into a frenzy: gnashing teeth, clubs against thick skulls, hurled dolomite stones, the new, "invading" Homo sapiens versus the old beetle-brows. It was breathtaking to envision sweeping hordes moving from deepest Africa up through the Mediterranean, trampling mud huts, raiding stocks of food, stealing caves, conquering and killing populations of unevolved *erectus* and Neandertals.

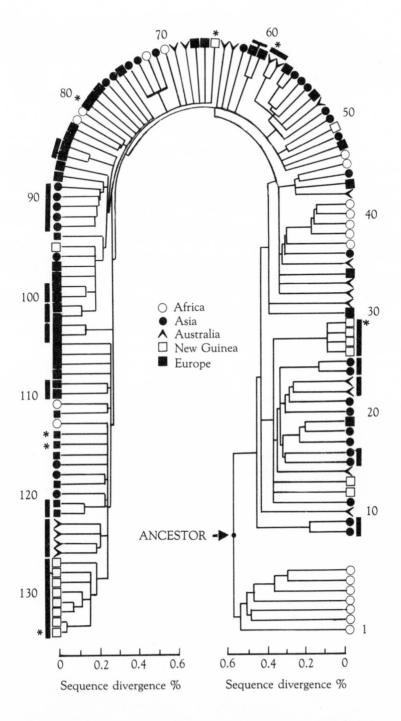

Genealogical tree for 134 types of human mitochondrial DNA

Although the geneticists, gentle in their lab coats, weren't trying to project any such violent scenario, the words meant to some paleoanthropologists that at one time in the past a population of Africans had taken over the world by raising something on the order of a Pleistocene holocaust. The most "rational, common-sense interpretation" of the DNA hypothesis is that the fanning out of modern humans and their replacement of those older and more primitive populations entrenched in Europe, Asia, and elsewhere in Africa involved "turmoil and bloodshed," says Harvard's Ernst Mayr, one of the century's great evolutionary biologists and the man who engineered the reclassification of Java and Peking men as *Homo erectus*. Mayr, who has been described as a final court of opinion in paleoanthropology, found aspects of the *Nature* piece "staggering."

Guaranteed to raise paleoanthropology's collective blood pressure was the assertion that there had been little or no intermixing between the African "invaders" and the Asian populations. Even the other Garden-of-Edeners accepted *some* reproductive exchange. The geneticists took the extreme view that there was perhaps *none*. That was what caused talk of a holocaust: how could there have been no sexual intermixing unless there had been outright warfare and annihilation of the Asian *erectus*-like people? But that was just a side issue. The real poignancy of what Cann, Stoneking, and Wilson were implying was that they were ready to dismiss a thick branch—or better put, part of the very trunk—of certain time-honored human trees. They were saying that populations in Asia were not only overwhelmed but contributed virtually no genes to modern humanity.

"We see no indication of any deep lineage in Asia," was how Wilson later put it.

The European Neandertals were implicitly disposed of, but that wasn't totally surprising. Now the geneticists were getting their turn to look back at the crucial Asian specimens, however, and what they had to say, in the *Nature* report, must have caused Weidenreich's own bones to turn in the grave. The geneticists themselves emphasized the original, founding role of *erectus* in human evolution, but they were referring, as I said, to an African *erectus*—not Java or Peking Man. Theirs was a beetle-brow south of the Sahara who turned into modern form before the Asian *erectus* could do the same, and then overtook all the Java and Peking descendants, who left not a trace.

"The greatest divergences within clusters specific to non-African parts of the World correspond to times of only 90,000–180,000 years," the geneticists said. "This might imply that the early Asian *Homo* (such as Java man and Peking man) contributed no surviving mtDNA lineages to the gene pool of our species. Consistent with this implication are features, found recently in the skeletons of the ancient Asian forms, that make it unlikely that Asian *erectus* was ancestral to *Homo sapiens*."

They were trying to get rid of Dubois's *erectus*! They were saying that Peking Man too was pretty meaningless, despite fifty years of uninterrupted hype. They were knocking from the human lineage a specimen so revered by Chinese as a sacred ancestor that even Mao and Henry Kissinger were peripherally involved in trying to locate the Zhoukoudian fossils when they were lost during the last world war. They were screwing up the dioramas in a hundred museums!

It was one thing to discard the klutzing, lame-brain Neandertals of Europe, who never had been totally accepted as direct human ancestors and who, Cann said, showed no evidence of contributing maternal lineages to modern Europeans, but the geneticists were also strongly implying that all those furry cavemen who had operated with cunning about the caves of Asia—for millennia—had no direct bearing on human evolution either. Those precious fossils from the limestone near Trinil and Beijing, the most ancient form of true man known to science for much of this century—the skilled hunters and firemakers and cannibals, prototype, along with Neandertals, for a thousand cartoons and the unilinear charts of human development—had sown no permanent oats. Their heyday was now meaningless history.

It was Maalox time for textbook editors, not to mention the encyclopedias. In place of the simple line charts—Java Man → Peking → Neandertal → *Homo sapiens*—was now a genealogical tree that resembled a huge, encrypted horseshoe. There were 134 little twigs to represent the genetic sampling, with two major branches connecting to the trunk, which represented the mito-

chondrial mother. It was a maze. Ancestral lines like a diagram for
an electric plant. Next to it was a long and undecipherable block
of code numbers listing what they called "polymorphic restriction
sites." It looked like something only a computer could read.

While one branch of the tree led back solely to Africans,
the second branch also included Africans, along with Asians,
Australians, New Guineans, and Europeans. That was a fairly
strong suggestion that Africa was the origin. "This inference
comes from the observation that one of the two primary branches
leads exclusively to African mtDNAs while the second primary
branch also leads to African mtDNA," was the way Cann, Stoneking, and Wilson put it. Moreover, they said, the Africans dis-
played the most diverse DNA types. (Again, the longer time a
lineage is around, the more mutations it collects, and thus the
more diversity within such a population.) It was a fairly compelling
indication that Africans were the originals, the oldest. Their divergence was 0.47 percent. Asia was a distant second, with 0.35.
Those two were followed by Australian Aborigines, New Guineans, and finally Caucasians, who weighed in at only 0.23 percent.

This was what Ernst Mayr found staggering: the Aborigines
of Australia appeared genetically closer to Europeans, in some
cases, than they did to Africans!

But then, the geneticists were also saying that there had
been no distinct races until after much of the world was settled,
so back then, who knew what a European or Australian looked
like? They hadn't settled down to form races yet. And we shouldn't
get away from the more immediate issue of intermixing. "If there
was hybridization between the resident archaic forms in Asia and
anatomically modern forms emerging from Africa, we should expect to find extremely divergent types of mtDNA in present-day
Asians, more divergent than any mtDNA found in Africa," said
the geneticists. "There is no evidence for these types of mtDNA
among the Asians studied."

Africa, it was stressed again and again, "is a likely source
of the human mitochondrial gene pool."

Archeological evidence, they said, was also pointing to an
African origin for mankind's shared grandmother.

After all, blades were in common use there long before they replaced the flake tools in Asia and Europe.

For the paleoanthropologist, it was not going to be a relaxed day of ham and football. To recap where we were now: The man-apes known as *Australopithecus* still stood as the supposed ancestors the farthest distance back (best remembered as the Taung child but best visualized as an erect chimpanzee), and of course there was still *Homo erectus* in our more immediate lineage, though minus the Java and Peking specimens that once stood as the very symbols of fossilized ape-men. It was still true that *erectus* had spread around the world about a million years ago, but only a small, isolated portion of those primitive humans—presumably a type developed in Africa alone—had evolved into the *Homo sapiens* who then took over the globe, replacing the *erectus* from everywhere else. Woe to those who thought modern man had started 40,000 years ago, as a Cro-Magnon in France. Woe too to the religious fundamentalists, like the followers of Rev. Jerry Falwell, who believed man was created in 4004 B.C.

One of the geneticists, Rebecca Cann, could barely contain her disdain for popular genealogy. "Dioramas in a natural history museum can be counted on," noted Cann, "to show some primitive ape-human (*Australopithecus*) emerging into some early species of our own genus *Homo*. Then *Homo erectus* evolves into *Homo sapiens*, along with mortgages, star wars, and ice cream. Such schemes imply a genetic continuity in space and time that contains more fantasy than Spiderman's best escapades."

The geneticists weren't mincing many words. It was a preemptive strike. They knew they were in for heat. Already a paleoanthropologist from Michigan had been grumbling very loudly and was mounting a major attack. Cann and Allan Wilson, who had fought long and hard with old-school paleoanthropologists, had the opportunity now to get their digs in. Both were calling for increased interaction between paleoanthropologists and geneticists—peace—at the same time they were firing off subtle and sometimes not-so-subtle salvos. The geneticists were in effect pronouncing themselves as the best hope for settling age-old disputes and finding the real answers.

Fossils would have to take a back seat.

"Besides the likelihood that the most sought after bones never will materialize," wrote Cann, "there is the problem of properly identifying the fossils that *are* found. It is hard to know, by bone shape alone, whether a fossil represents a species already identified or whether it is different enough to represent a species of its own. Then there is the difficulty of knowing whether the fossil was left by a human ancestor or by a related primate that became extinct. All in all, it is too much to hope that the trickle of bones from the fossil beds of eastern Africa will, in itself, provide a clear picture of human evolution any time soon."

Elsewhere she added pointedly: "As molecular biologists, we at least knew that the genes available from present-day specimens came from some ancestor. In contrast, paleontologists can never be certain that a given fossil has left descendants."

Wilson was blunter still. "Some people don't like our conclusions, but I expect they will be proved wrong again." The old-school paleoanthropologists, relying as they did on analyzing fossil fragments and physical features (the practice known as "morphology"), would probably have to be dragged by their own noses, but they would come around to his view. He'd seen it before, and he wasn't happy at all with paleoanthropology. Back when he was studying man's relationship with primates, the paleoanthropologists had badly shunned him. He wasn't the sort to forgive and forget. He had once said his group had developed "a set of data that's much larger than I think has ever existed. The morphologists can't come close to us." Now there was more data still. And more sophistication. The paleoanthropologists had been left in the dust.

It certainly sounded like war was developing between the fossils and genes. It was a war that had been a long time in the making. Of all the nerve: upstarts from Berkeley and Hawaii deciding they will provide the final answers not with an ossified skull but with genetic mutations. The big paleoanthropological news of the year was not made by a suntanned, dust-laden paleontologist in the outback of the Fayum Desert, not by the unearthing of a skeleton caked in breccia or kept preserved in a prehistoric bog, but rather by those stiff and haughty lab coats.

A new age was clearly dawning. While street sweepers were

piling kazoos and party hats in the gutters of Times Square, the geneticists had completed the big play, opening the new year with a discovery that—though not yet garnering the same level of publicity—rivaled the most celebrated of fossil discoveries.

Could it be true that the geneticists did hold the key to the past? Was Eve alive and well in the whir of a centrifuge?

4

THE LAB WAS at Berkeley, the biochemistry building at the very entrance of the campus, 15,000 miles from sub-Saharan Africa. Instead of dust devils or steaming primordial jungle, there was only white smoke spewing from a cylinder of nitrogen and the vague aroma of chlorides.

It was a clean, well-lighted place, but no spot for a picnic: pH papers, a jar of lizards, books with titles like *Of Urfs and Orfs*, refrigerators full of enzymes, and latex gloves on every counter, along with the crystals, powders, gels.

Another volume of notes was labeled "Lesions Found in Mummies." One of the geneticists looked at molecules scraped off old Egyptian remains. He was currently analyzing DNA from the remarkably preserved brain of a nearly 8,000-year-old man who had been found in a Florida bog. There was also a road map for Australia. And signs warning of radioactivity. The centrifuges were in their own separate little room. They looked like washing machines.

Although other laboratories had initiated the use of genetic information to track human evolution, and although the specific use of mitochondrial DNA from humans for this type of work had been pioneered by a biologist named Wesley Brown, it was in the mid-1960s, in this same building though on a different floor, that the search for a molecular Eve, in many ways, had its origin. The effort had started as a comparatively low-tech investiga-

tion comparing proteins from humans and apes in order to determine how close they were to each other. That work had been done
by Vincent M. Sarich, a tall and bearded biochemist with jolly,
mischievous eyes, who put one in mind of Ichabod Crane; and the
head of the lab, Allan Wilson, a white-haired native of New Zealand who was much shorter than the gangling Sarich and much
less outgoing, even dour. Now in his fifties, Wilson was an uneasy
man who often was described as a recluse. Although he would
embark upon a course of increasingly spectacular speculation—to
the point of stunning fellow evolutionists—Wilson was a highly
regarded biochemist, and for good reason, by the appearance of
things. He needed a secretary to keep track of all his journal
articles, and he hid in a side office amid a mountain of so many
papers and books that it was hard to find a place to set anything
down.

If two men could hardly have been more different physically, and if Sarich was often overtly aggressive while Wilson spoke
in a rather shy and fragile voice, both were extremely innovative
thinkers, "big-science" guys, and both were always being accused
of arrogance. They also shared what seemed like a hatred of fossils.

Which was odd in that Sarich, besides his credentials in
biochemistry, was also a doctor of anthropology. The son of a bus
driver in the Chicago area, he had always been the precocious
kind, skipping a grade in grammar school, the type of kid who had
a chemistry lab in the basement. He attended Illinois Tech, where
he majored in chemistry, but chemistry bored him. After a stint
in the Marine reserves, he found himself enrolled in Berkeley's
anthropology department. Soon he was heading for his doctorate.
His wife Jorjan was an anthropology major and it was through her
studies that Sarich's new interest had been born. He read whatever
he could get his big hands on, which at Berkeley was a considerable
mass of material even in those days. One of the nation's foremost
paleoanthropologists, Sherwood Washburn, was there, as was fossil hunter Clark Howell (at the time a visiting professor), and
Berkeley was fast becoming a citadel of paleoanthropology. In the
cellar of the big old Life Science Building were cabinets of animal
and human bones, and there was a 2,400-year-old sarcophagus in
the expanding museum at Kroeber Hall.

Sarich's direct involvement in molecular evolution began with a seminar conducted by both Washburn and Howell in the fall of 1964. He volunteered to do a presentation on systematic serology and how it might apply to primates. Much of the work that would follow harkened back to George H. F. Nuttall, a turn-of-the-century biologist from Cambridge who had speculated that the chemistry of blood proteins could be analyzed to gauge the genetic similarity between various animals. Nuttall's groundbreaking work was followed up by Morris Goodman at Wayne State University in Michigan, who used Nuttall's approach on serum proteins to show the relatedness of humans and African apes. It was something that took an immediate intellectual hold on Sarich, who was egged on by his mentor, Sherwood Washburn.

"Nuttall's method was immunological," recalled Washburn. "If blood serum from an animal is injected into an experimental animal, the experimental animal will manufacture antibodies against proteins in the foreign serum. If serum from the experimental animal is added to serum from a third animal, the antibodies will combine with similar proteins in that serum to form a precipitate. The stronger the precipitation reaction, the closer the relation of the first animal to the third."

The protein structures of blood were, in a way, a crude reflection and manifestation of the genetic material, for DNA's chief function is to code for the cellular production of amino acids for those building blocks of the body itself—proteins. It was a couple steps removed from DNA, like knowing a house from the bricks you see instead of the intricacies of the actual blueprint itself.

Washburn (who like Sarich and Wilson after him was dissatisfied with the sweeping conclusions being drawn from mere fragments of fossil) had invited Goodman to a conference in 1962 that was held in a lavish twelfth-century castle about fifty miles from Vienna. That added a little mystery and maybe even something sinister to it all, and the symposium became a landmark in the development of the theory of molecular evolution.

"I had established by 1962 that the African apes and humans were genetically closer to one another than to other primates, and I proposed then that this could be reflected in a classification which removed the African apes from the *Pongidae* and placed them in the family *Hominidae*," Goodman told me. "Moreover, I also pointed out that the three genera—*Homo,* chimp, and gorilla—diverged from one another to the same extent. The morphological view generally places the chimp and gorilla much closer together than to humans. But the genetic data as it's accumulated really shows something closer to a three-way split."

So primate evolution appeared to be shaping up like this: first you had ratlike, insect-eating little mammals that had begun clinging to trees (picture the tree shrew); next something like a bushbaby or lemur, with eyes moving to the front of the head, a pre-monkey or "prosimian"; then, from that rootstock, a whole cacophony of monkeys in South America (the "New World"), along with Europe, Asia, and Africa (the "Old World,") once the continents floated apart. As evolution supposedly proceeded, there had been another major split, this time between monkeys and creatures developing into the large, tailless, semi-erect apes. After millions more years there was yet another branching off that went in three directions: one toward the chimpanzee, one toward the chest-thumping gorilla, and another toward hominid man-apes and finally the furless *erectus* and *Homo sapiens sapiens,* who were all but nude.

While Berkeley's Sherwood Washburn was one of the few in his field who believed in a recent ape-human split, and was greatly impressed with what Goodman had to say about how close the species were, he was always on the lookout for more detailed information about the relationship between man and chimp. For instance, no one had determined with molecules when exactly the evolutionary branching had occurred. Though all in the mood himself to accept tree-swinging (or "brachiating") African apes as very close to our own ancestry, he was in search of a secret weapon to prove his point.

This was where, in 1964, at another seminar Washburn was conducting, tall, mischievous Vince Sarich, a student of Washburn's, came in. Sarich almost immediately set his sights on the ramifications and further uses of immunological techniques. His anthropological antenna was humming. Another member of the seminar was a student who happened to have worked in Goodman's lab at Wayne State. Sarich read through all the pertinent molecular literature—which back then took only a couple weeks—and made his presentation. It was so impressive to Washburn that the anthropology professor suggested the topic might make for an interesting dissertation. (Eventually Sarich would be identified as a member of both the anthropology and biochemistry departments, the beginning of a profound linkage between the two disciplines.)

Sarich showed the paper to Allan Wilson, who he had heard of, in fact, during a course in biochemistry. Wilson was in his first year as a faculty member in the department, his doctorate earned in biochemistry at Berkeley after studying zoology and chemistry at Otago University in Dunedin, New Zealand, and getting a master's in zoology from Washington State University. His postdoctoral studies had also been in biochemistry, this time at Brandeis University. Wilson had long held an interest in both evolution and biochemistry; one day he would be described as the man who brought molecular evolution to the human level. When he and Sarich met, he had been peering at the enzymes of various organisms—birds, chimps, ruminants—with his own immunological technique, which he picked up at Brandeis and called "microcomplement fixation."

Sarich and Wilson were soon to join forces. In these days before mitochondrial DNA, they were going to rely on the blood protein albumin for evolutionary studies. "At Brandeis, microcomplement fixation was shown to distinguish between human and chimp albumin," recalls Wilson. "So we could *measure* the chimp-human distance." It all came down to the protein differences between various animals again. Already Wilson had published one paper on molecular evolution, but it had been in a journal Vince Sarich had not seen.

If the paper had been of no immediate importance to Sarich and Washburn (nor one that greatly stirred the world of evolution as a whole), they were nonetheless impressed by Wilson's obvious professionalism. He seemed calm, accurate, "death on details and good lab work," in Sarich's recollection, possessing many of the characteristics, adds Washburn, that "the public thinks scientists should have—not flourishing a war cry for something or another." At the time he was quiet—aloof to the point of nearly recoiling from people—but very strong-willed and an indefatigable worker. Besides the immunological technique, Wilson had something else neither Sarich nor Washburn did: a lab facility.

The lab was just getting started down near the corner of Oxford and Hearst. Sarich began helping Wilson arrange the place, collecting glassware and searching for other bare essentials. They would use Wilson's fixation method to determine the genetic distances between monkeys, apes, and human beings.

So by 1965, in his own diffident and aloof way, Wilson, with some help from Sarich, was developing what would become a whole suite of laboratory rooms—the rooms where, two floors lower and two decades later, a swarm of postdoctoral students would be rushing around with enormously complex techniques, studying mummies, tissue from a frozen 40,000-year-old mammoth discovered in Siberia, and bits of muscle from an extinct, zebralike animal, the quagga, scraped from an old museum mount.

It was also where Eve would make her dramatic appearance.

Sarich's technique involved going to a blood bank to have his own blood drawn for a sample, extracting albumin from it, and injecting the albumin into rabbits. Since the human albumin was a foreign substance to the rabbits, they raised antibodies to it. The rabbits were only peripheral to the actual experiment, serving in effect as antibody factories. The antibodies, in their turn, were but a molecular yardstick.

Sarich bled the rabbits and reacted the antibodies in a solution with more of his own albumin. The mixture caused a precipitate he could see. The stronger the reaction, the more precipitation. The antibodies were not only a yardstick but also something like litmus paper. Sarich then gathered samples of ape and monkey blood, purifying albumin from them too so he could see how *they* would react with antibodies that were specific to human albumin. The ape samples were sent by Morris Goodman, the monkey blood was drawn by the Davis Primate Center. The antibodies created to defend the rabbit against human albumin would react most strongly—forming the thickest precipitate— when it met up with more human albumin. The next strongest reaction would occur when the human-oriented antibodies were reacted with the albumin of whatever animal bore the greatest molecular resemblance to human beings.

"Literally, if instead of my albumin I stuck in a monkey's albumin, there's sufficient similarity so that the rabbit antibodies would recognize the similarities between my albumin and the monkey's albumin and precipitate those as well," explains Sarich. "But since it is only similarity—not identity—you would get less. So the measure of difference between species is the degree of precipitation."

It sounded like a roundabout way of comparison but in reality was about as straightforward as molecular science gets. Proteins evolve over time just like the organisms of which they are a part. By gauging the precipitation, they could measure the difference—the genetic distance between two animals—in ratios, percentages, or other numbers. The further away an animal was to man, the higher was the index of dissimilarity.

Over the course of expanded experimentation Sarich and Wilson noted that chimp and gorilla albumins reacted the strongest with human-specific antibodies, followed by Asian apes like the orangutan, various types of monkey, and at the lower end of the spectrum animals such as the cow or pig. So there was a scale, and it could be expressed quantitatively. "Consider the precipitate from human albumin to be 100 percent," explains Sarich. "A chimp would be about 96 percent as much precipitant, an Old World monkey like a baboon 70 to 75 percent, a New World monkey like the spider monkey 60 percent, a lemur 35 to 40 percent, a dog 25 percent, and a kangaroo only 10 percent."

That gave a fairly well-defined idea of diversity. If the scale of dissimilarity was overlaid on a chart of the chronological sequence of evolution culled from the fossil record—which placed the pre-monkeys, for example, at somewhere around 60 million years ago—one could superimpose the genetic differences between various animals and get an idea of when they branched off from one another.

The key question seemed to be whether changes in protein molecules like albumin—the mutations that accumulate through the ages, as evolution takes place—built up regularly in the molecules, like sand in an hourglass, at the same rate in one species as another, or whether they happened irregularly. If the rate of change was highly *irregular*—in fits and starts, as was generally perceived—that would allow for the possibility that the small difference Sarich, Wilson, and Goodman had noticed between man and apes was not because the two had split from each other recently, but because molecular change in humans had instead encountered a slowdown that only made it *appear* like the evolutionary split had been recent. If the sand in an hourglass suddenly slows down to a trickle, anyone judging time by it will think less time has transpired than is actually the case.

In other words, there was the chance that apes and man had diverged from each other as long ago as the mainstream paleoanthropologists thought—15 or 20 or even 30 million years ago—but that their blood proteins had not accumulated very many differences over the course of millions of years because of a slowdown, giving the false impression of a recent divergence from one another. Unless Sarich and Wilson could establish the fact that changes occurred on a regular basis, there was always the chance that the ape-human split, contrary to what they and Washburn thought, was very ancient indeed.

Sarich and Wilson knew, from a paper published in 1962, that studies of hemoglobin by Linus Pauling, the Nobel prizewinner at the California Institute of Technology, and Emile Zuckerkandl, who was in Montpellier, France, indicated that molecular changes probably did occur on a regular basis—like a biological hourglass. Pauling and Zuckerkandl had more than hinted at this regularity when they stated that "it is possible to evaluate roughly and tentatively the time that has elapsed since any two of the hemoglobin chains in a given species . . . diverged from a common chain ancestor." Count up the mutations, look at the amino acid differences between species, and you may have an idea of time. The two scientists went further in a 1965 paper, saying it was possible that "changes in amino acid sequence that are observed to be approximately proportional in number to evolutionary time" may have been the result of minor mutations that did not tremendously change the function of the organism but did occur with that all-important regularity. "There may thus exist," said Pauling and

Zuckerkandl, "a molecular evolutionary clock."

Thus the idea of a built-in evolutionary clock had been floating around the intellectual ether for a few years, and Sarich had certainly found it intriguing. It seemed so elegant a concept, and so simple: the longer two species had been separated, the more ticks had tocked.

But first, of course, there was the matter of checking for the clock's regularity. If the molecular clock ran fast and slow, depending on the species, you couldn't so easily project back into the past. The paleoanthropologists, believing that the ape-human branching was long ago, tended to think that just such a slowdown had occurred, skewering the interpretations of time. If lower primates like monkeys had fast clocks and higher primates like apes and man slower clocks, it would only *look* like man and apes had a recent ancestry when, in fact, more time had elapsed than the slower clock had recorded.

"It turned out," noted Sarich, "that when we measured the distance from man, chimp, gibbon, baboon, squirrel monkey, and so forth, there *wasn't* any slowdown. And if there wasn't a slowdown, then there was a clock!"

Tick-tock. They had found the timepiece. It was in albumin, that blood protein, and it functioned, they were convinced, with crucial regularity. Sarich and Wilson learned that descendants of various primate lineages had collected about the same number of protein mutations over similar periods of time. Now they could use the scale of molecular differences to project backward in time. All they needed was a little calibration.

The oldest primates—prosimians much like the lemur, a raccoonish little tree dweller from Madagascar—were in existence around 60 million years ago. If man and the apes had split and gone their separate ways 20 million years ago, as so many paleontologists believed (actually 15 to 30 million), then the difference between humans and chimps should be about one third of the molecular difference between humans and lemurs. Instead of being a third, however, it was more like one *fifteenth*. A fifteenth of 60 to 75 million is 4 million years and, with variations taken into account, the figure can be calculated at about 5 or even 6 million years.

As Wilson sums it up (with a little characteristic sarcasm at the end), "The molecular evidence indicated that the ape-human relationship was extremely close, with a five-million-year separation rather than a thirty-million-year separation. And given that point of view, it required a reinterpretation of those fossils, and it took something like twenty years for that revolution to happen."

The "revolution," for all intents and purposes, began in 1967, when the two young and supremely confident Berkeley scientists had co-authored a paper entitled "Immunological Time Scale for Hominid Evolution." The paper was printed in another prestigious journal, *Science,* and what it had to say, by the standards of that time, was as iconoclastic as the later *Nature* paper. As we have just seen, back then zoologists believed that apes and humans diverged from each other—that is, split from some kind of common animal ancestor—a very long time ago, perhaps 15 to 30 million years in the past.

The time range was important because it seemed necessary to keep apes in the long-ago past. Twenty million years, for example, was a safe and distant 200,000 centuries. In the mind of the average person, chimps and gorillas were naught more than overgrown monkeys. While it was generally agreed that the African apes (which include chimps and gorillas) were our closest relatives in the animal world, they were just that: animals. If, in reality, they were quite a bit more evolved than a spindly, long-tailed, chittering monkey, still, nothing could negate the fact that apes too were kept in a cage or lived around jungles. Certainly gorillas and chimps were closer to each other than they were to *man.* They both had fur. They had no words. They walked on all fours, using their beastly knuckles.

The experiments Sarich and Wilson had conducted purported to show, contrary to general belief, that chimps and gorillas are as close to humans as they are to each other. Their albumin was so similar, said Sarich and Wilson, that they were as close to each other—man, chimp, and gorilla—as a zebra was to a horse! From one molecular standpoint, humans and the chimpanzee were 99 percent alike, Wilson alleged. Gorillas were only slightly less close than the chimpanzee. All three had shared a common ancestor much more recently than most everyone supposed. "Our calculations led to the suggestion that, if man and Old World monkeys last shared a common ancestor 30 million years ago, then man and African apes shared a common ancestor 5 million years ago, that is, in the Pliocene era," Wilson and Sarich wrote. It was no minor claim. According to these scientists, the chimp was our brother, not just a distant cousin.

This bordered on the heretical, radical even to those who subscribed to evolution unquestioningly. Sarich and Wilson were saying that man and the apes shared an ancestry that was much more recent, shockingly more recent, than the fossil evidence had indicated—three, four, or even six times more recent than various paleoanthropologists had proposed.

Some of the fossil evidence indicated that man had a baboon-sized ancestor called *Ramapithecus* 7 to 15 million years ago. A big deal to paleoanthropologists, *Ramapithecus* was supposedly a "pre-hominid" that came before australopithecines such as the Taung child, a bridge between apes and man-apes. But according to Sarich and Wilson, *Ramapithecus* could not possibly have been an early hominid—a direct ancestor of man, coming after the ape-human split—if man had diverged from apes only 5 million years ago. It would have existed *before* there was an ancestral split between apes and hominids. And so the beloved *Ramapithecus* was demoted by biochemistry back into being a simple ape instead of a man-ape of any kind.

Vince Sarich didn't take a bullhorn to the streets, not yet, but he did analyze just about anything available: langurs, vervet monkeys, pygmy chimpanzees. He even analyzed the tarsius, a little prosimian that looks like a bat and clings to Indonesian trees. The molecular evidence coming not just from Berkeley but a burgeoning number of other labs was nothing less than peculiar. It would one day be reported, for example, that a pig was closer to a whale than it was to a horse, a pigeon or penguin was more closely related to a turtle than a turtle was to a fellow reptile like a snake, man was closer to mice than rabbits, and while the giant panda was a bear, the lesser panda was more of a raccoon.

As usual, Wilson kept quiet about the original findings. He preferred to avoid direct interaction with the public, instead making big declarations at insulated scientific seminars or in technical papers that were excruciatingly detailed and yet jumpy and scattered, devoid of a comprehensive framework. Most of the time he was not much for explaining things. Though the fray would become such that he wouldn't be able to resist a few zingers, Wilson preferred to wage his campaigns in shrewdly crafted journal articles, and then cloister himself in his office, with an almost religious reverence for his bread and butter: the molecule.

While Wilson kept close to the shadows, Sarich couldn't contain himself; he all but declared the old stones-and-bones school of thought dead. It was the beginning of a battle that is still heating up today. What really thrilled Sarich was his belief that he was no longer beholden to fossils. Year by year, as more results came in, as albumin continued to look like an evolutionary clock, Sarich grew increasingly bold. "We didn't have to talk to any paleontologists, we didn't have to do anything but get blood samples," he still says gleefully.

"Allan and Vince are utterly different people temperamentally," says Washburn with nearly a sigh. "Wilson is calm, accurate, always had very carefully trained students working with him, and Vince is much more likely to go off on a tangent. So when it came to selling the notion that man was particularly closely related to apes rather than to monkeys and so on, Vince was much more important than Wilson in the sales line."

But what Sarich was out there selling, added Washburn, was indeed a very important idea. "The capacity of molecular taxonomy to define the relations among primates is perhaps the most important development in the study of human evolution over the past several decades," he said. Others would one day compare it, with distinct hyperbole, to the pronouncements of Darwin.

Whatever else they were, the findings were certainly radical, nearly a manifestation of Berkeley's overall, freewheeling, rebellious spirit. It was the sixties, the Free Speech movement. And Berkeley was at the eye of the storm. If the biochemistry department was thankfully on the other side of an imaginary line that separated it from social sciences, across the entire city were the stereos blaring revolution, the heavy scent of marijuana, and the ultimately irrelevant love beads and odd ways of shaking hands. Hippies rolled joints on University Avenue, but it was getting hard to tell them from everyone else, for the students too had taken to tattered jeans and shoulder-length hair. The earnest, upraised classroom hand, anxious to please a professor, had been replaced by a militant fist or the languorous peace sign.

The university itself was steadily gaining on prestigious schools like Yale, Stanford, and Harvard, but that was partly obscured by the uproar of the 1960s and also by the school's own reputation for exotica: high-energy physics, high-energy biology, the flashy stuff Stanford often avoided in favor of workaday drudgeries that brought practical applications out of new discoveries. Its obvious academic attributes notwithstanding—and these were becoming fairly immense—the city of Berkeley, getting back to the area's general ambience, was on the way to developing the reputation of a "flaky" place. One day it would be mocked as "Berserkely."

The paleontologists and paleoanthropologists did not immediately attack the "radical" molecular claims coming from "Berserkely." In a way, that stung the most. "Mostly there wasn't any reaction," says Sarich. "It wasn't taken very seriously. You had to realize what me and Allan were claiming: we didn't need paleontologists. We could figure all this out ourselves. These were questions that had bedeviled people for generations—more than a hundred years—and we were saying, 'Here's the answer to it.' We were challenging paleontologists and also biochemists by using a technique that wasn't in good favor with them, and also evolutionary biologists by suggesting there was an aspect of evolution that occurred in a clocklike fashion. They had convinced themselves that that was impossible. And we were nobody. It was out of the blue and challenged everybody. So it didn't ever get a really intelligent critique. It was out of line and so it was basically ridiculed or ignored."

Wilson was also irritated. He wasn't invited to anthropology conferences, which should have been anxious to hear what he had to say. He too felt the work was being ridiculed. Certainly no one was looking upon him as the new Darwin nor Sarich as a reincarnated Huxley. They were more like molecular Moonies. Wilson resented anthropologists and their fossils because, he said years later, "their science hasn't been a quantitative one and an objective one." The silence from paleoanthropology was maddening. "They functioned as if we didn't exist," complained Wilson, echoing Sarich.

When a voice was raised up, it was often indeed the voice of ridicule. John Buettner-Janusch, an anthropologist who had indirectly linked Morris Goodman up with Washburn to begin with, and had provided Goodman with some of his own ape samples, was an early pioneer of the molecular outlook. ("The era of the molecule, the protein molecule, is upon us," announced Buettner-Janusch, then an associate professor at Yale, a year before Wilson and Sarich published their first joint paper. "Anthropology, as it attempts to reconstruct the phylogeny of man and his fellow members of the order Primates, must take cognizance of molecules.")

Once Buettner-Janusch heard what Wilson and Sarich were saying, however, his tone turned to vitriol. He didn't like the assumption of a uniform mutation rate. He accused them of "careless and thoughtless statements." With dripping sarcasm he told one audience: "No fuss, no muss, no dishpan hands. Just throw some proteins into the laboratory apparatus, shake them up, and bingo!—we have an answer to questions that have puzzled us for at least three generations."

Buettner-Janusch was one anthropologist who played for keeps. He was later convicted of manufacturing drugs in his lab at New York University and, out of vengeance while he was on parole, he mailed valentine boxes of poisoned candy to the home of the federal judge, Charles L. Brieant, who had sentenced him in 1980. The judge's wife was hospitalized after eating some of the chocolates, which had been laced with a potentially lethal dose of pilocarpine hydrochloride. Also made ill were the wife and daughter of another target of his revenge, Dr. J. Bolling Sullivan, an associate professor of biochemistry at Duke. Meanwhile the NYU professor, who had once headed the school's anthropology department, ended up getting forty years and writing his papers from prison.

A tough customer, Buettner-Janusch, and if he didn't have much credibility in the end, the other paleoanthropologists, some of them equally vitriolic, certainly did. They were astute men who had worked hard in the caves and rock shelters, and they weren't going to be deposed by a couple of molecular biologists. Their own science—comparing all those morphological traits—was in many ways every bit as complex and challenging as biochemistry. And they were hardly going to concede that the primary tool of their trade, fossils, was now all but irrelevant. The granddad of fossil hunters, Louis Leakey, threw his weight against the molecular conclusions, arguing that they were out of accord with the fossil record. "The average reaction of the paleoanthropologists was that it was all nonsense," confirms Washburn, whose contacts in the field were extensive. "People like Elwyn Simons were very upset. So was David Pilbeam."

Simons and Pilbeam, then both at Yale, were the champions of *Ramapithecus,* which of course Sarich and Wilson had demoted to irrelevance. The molecular data meant the Yale group was dead wrong. *Ramapithecus,* represented by fragments of jaw found in northern India, wasn't a hominid, it was just some kind of an extinct ape. "One no longer has the option of considering a fossil specimen older than about eight million years a hominid *no matter what it looks like,"* Sarich said in 1971, absolutely enraging the fossil hunters.

Venerable old Sherwood Washburn, watching what he had helped to turn loose, was now cringing a little. "I *begged* Vince to take that out of the paper—literally begged him—on the grounds that [it was] just going to annoy people," he recalls. "Vince is a tough guy. He wasn't going to change it. Vince enjoys intellectual warfare."

It was biochemists versus old-school anthropology, Berkeley versus Yale, Berkeley versus everyone. The only thing missing was a UN peacekeeping force. Sarich was forging ahead relentlessly. Whoever wasn't on their side was the enemy. Sarich and Wilson even began sniping at Morris Goodman, who had provided their very first ape samples and whose early serum experiments had set the way. They were disgusted because Goodman believed, contrary to their findings, that there *had* been a molecular slowdown in man. "Why does he say it?" Wilson asked Roger Lewin, a writer at *Science*, with obvious pique. "He says it because he is afraid to take on the paleoanthropologists."

Whenever the maelstrom showed signs of abating, Sarich had another taunt to throw in. He was distressed that the fossil evidence was given priority over his biochemical results simply because it had been around longer and had been the only means of tracking down ancestors for a long period of time. "With the development of biochemical evidence, that situation is changing," he pronounced in the 1973 *Yearbook of Physical Anthropology*. In a statement fore-echoing what Rebecca Cann would one day say about DNA, he added bitingly, "One might, with some exaggeration, now epitomize the contrasts in the current picture by arguing that the biochemist knows his molecules have ancestors, while the paleontologist can only hope that his fossils have descendants."

It was a dig at all the precious fossils that were really only extinct, dead-end ape-men. The paleoanthropologists were blinded by prevailing orthodoxy, Sarich argued, and as for Morris Goodman, even though the Wayne State biologist had "laid the modern foundation for meaningful integration of macromolecular data into the development of evolutionary scenarios," he had also sold out, conceding to the paleontologists "their coveted arbitrators mantle." The main problem, as Sarich saw it, was that "paleontologists have arrogated unto themselves the mantle of final arbiters of our understanding of the evolutionary process—in effect, if we cannot see it in the fossil record, it never really happened."

Although Elwyn Simons, the Yale expert on primates (now at Duke University's primate center) and proponent of *Ramapithecus,* was fully taken aback, he and many others in the field would eventually come to accept at least some of what Wilson and Sarich were claiming. *Ramapithecus,* based on molecular and new fossil evidence, is now dismissed from man's lineage, with the date of ape-human divergence set closer to what the Berkeley group had said all along. But Simons, who had personally resurrected the *Ramapithecus* fossil, never would get over their approach to the issue. "Sarich was *extremely* swaggering and arrogant and he was not like most people I've known in science," he says. "He came on massively strong that he had to be right and people that studied fossils were a bunch of fools. He radiated that by everything he said and did. He did say things to the effect that biochemistry was the only way to do science and other people were crazy."

So, from the sixties into the 1970s, even before the real heavy genetics was starting up in Wilson's lab—long before his carefully trained workers, laboring with fancy new enzymes, found the mitochondrial Eve—a new way had been paved to search for man's roots. A way had also been paved for a riot of controversy.

5

WHILE VINCENT SARICH continued his side of the debate and expanded upon his testing, heading to New Guinea to collect bats and gathering blood from harbor seals, walruses, mongooses, and even something called the binturong, a young Coloradan named Wesley Brown was just finding himself as a biologist, his interest also honing upon the animal world, but from the perspective of mitochondrial DNA instead of the cruder measure of albumin.

Where Sarich and Wilson had judged genetic differences between animals by the precipitation of proteins, which are designed by DNA, Brown was going to compare some of the intricacies of the DNA itself. Where the Berkeley scientists found a surprisingly close relationship between chimps and humans, Brown was going to confirm some of that work but take it a step further, comparing humans with other humans instead of with apes. And so where Wilson and Sarich had provided the general scheme of the house, Brown was eventually going to stop by their lab with the actual blueprints.

The blueprints provided by this tall Coloradan with sandy-brown hair would contain the first suggestion that mitochondrial DNA might reveal not only the relationships between various and sundry organisms, but could also be used to project backward and locate a small and ancient population that had founded the entire human race.

Wes Brown was no laboratory robot. He wasn't much into

the ivory-tower scene. And unlike some of his colleagues, he didn't act as if he alone held the key to human existence. There was a healthy lack of pretension when he viewed himself. He had been a "screw-up" at the University of Colorado, up there in Boulder near the foothills of the Flat Irons, partying hard and majoring in political science because it carried a smaller workload than just about anything else. Upon graduating he had no idea what he wanted to do. Boulder too was revving through the sixties. You were either straight or cool. Mostly you floundered. Wes Brown was a flounderer. After college he had simply packed a few bags and gone to Europe. For some reason—probably because he'd always liked animals—he found himself in Frankfurt reading biology textbooks.

So much for political science. Brown's fascination was the relatedness of living things. As a child he would sit around scribbling out his own little phylogenetic charts, linking fish with the whale because they looked alike and both swam in the ocean. It didn't matter that one was cold-blooded and the other a mammal. You had to start somewhere. He certainly was not at the point of being able to link whales to pigs. From that beginning so long ago he had leapt to DNA. He also jumped headlong to the matter of human evolution, which meant that besides a little partying, he liked to run through hoops of fire.

"I went over to Europe to grow up, and I did, and I came back, and by a great stroke of fortune I'd had general biology [at Boulder], two courses, and another in genetics, and I had the same professor for one of the two semesters of the general biology and evolution and genetics, and he was really good and I did well in those courses," recalls Brown. "When I got back, I went over to the Boulder campus to talk to that professor and told him I wanted to come back to grad school—in biology! He knew me and happened to be chairman of the graduate admissions committee for biology at the time, and he went to bat for me. So they admitted me and I took courses and did well and got a master's degree in evolutionary biology."

Brown had also taken some biochemistry, and when a Yale scholar passed through talking about isozymes, Brown really flashed on that. He had a keen interest in molecules, wanting to link them up with evolutionary research. "Too many people were looking at morphology, and some of what they'd come up with seemed farfetched," he says. "People had been looking at morphology for three hundred years and the picture hadn't been clarified all that much."

Brown went on to Arizona State University for a year because he was interested in lizards and snakes, which were plentiful in the nearby desert. But his next big stop was the California Institute of Technology, where they were looking at mitochondrial DNA from the standpoint of molecular biology.

Since 1850, scientists studying cells with light microscopes had noticed the presence, in those cells, of the small threads, rods, and granules that we today call mitochondria—the Greek is *mitos* for "thread" and *chondrion* for "granule." Upon further investigation, biologists of the last century were able to determine that these subcellular granules were quite similar in shape and size to bacteria.

For the first several decades of this century, scientists looking at these structures argued about their function. Some thought they were where protein was synthesized; others figured they served as genetic information centers. In 1932, a decade before Wes Brown was even born, a granular fraction was isolated from the liver of a guinea pig, and during the next two decades scientists got a closer look at the mitochondrion's wavy, convoluted membranes.

This was all occurring in that sappy stuff outside the cell's nucleus—the cytoplasm. That's where mitochondria collected, sometimes looking more like zebra-striped potatoes than threadlike granules, an organelle that ended up being the cell's energy facility, respiring and munching on bits of carbohydrates and fats. The main function is to generate power by combining oxygen with those food molecules. The energy of mitochondria is synthesized as the universal energy currency, ATP. Without these organelles, our biological processes would shut down. In each cell are about a thousand of them.

There were indications, years before Brown had become interested in the subcellular unit, that, like the cell's nucleus, where the vast number of inherited characteristics are stored, mitochondria too possessed certain of their own, independent genetic properties. That is, mitochondria—these power centers—had their own DNA. There was a mutation found in yeast, for example, that was cytoplasmically inherited—from outside the nucleus's chromosomes—and in 1964 DNA was found in yeast mitochondria. The mitochondrial DNA is a small, double-stranded molecule, appearing under the electron microscope like a blob with uneven contours that put one in mind of New Jersey. It is far less complex than nuclear DNA and many, many times smaller. It seems to evolve of its own accord, without getting reshuffled during reproduction—inherited solely from the mother.

When most people think of genes, they are thinking only of the nucleus, and for good reason. It is the nuclear DNA in a male's sperm that unites with that in the egg of a woman. There is a mixing together of genes like marbles in a tumbler. Characteristics from both parents are manifest in the baby, a unique human being with a mixture of the parents' genes.

Trying to sort through all the nuclear genes contributed from both parents is a trying, frustrating, and thus far incomplete task, but in the 1960s scientists suddenly had a second source of DNA to study—a different, simpler DNA—in the mitochondrion. When sperm enters the ovum, only nuclear DNA is believed to enter the egg in any appreciable quantity. The cytoplasm and its mitochondrial DNA, which has very little to do with most inherited features anyhow, can't fit in the sperm cell or, depending on whose version you buy, is unceremoniously cast off. In some cases sperm mitochondria might swell and disintegrate.

So it is that when a baby's cells form and duplicate, they contain nuclei possessing the genes of both the mother and father but mitochondria—those little potatoes—containing the genes only of the female parent. It has been that way since the beginning of man. Like the proteins, these genes were thought to mutate at a roughly constant rate over the ages, but to do so in a way that is relatively easy to decipher. The variations are from mutation, not recombination. And they happen as "errors" when DNA is copying itself. So, like proteins, the more mutations that separate two organisms, the longer they have been separated evolutionarily.

The mitochondrial DNA, as Brown saw it, "appears to proceed in a much simplified and more straightforward manner than the evolution of nuclear DNA."

The young, itinerant Brown, with circular, double-stranded molecules dancing in his head, soon found himself laboring in a lab run by Cal Tech chemist Jerome Vinograd. The lab had been working on the basic structure of mitochondrial DNA and investigation of the unusual genome had become an international investigation. A Dutch scientist was looking at mitochondrial DNA from chicken liver, while scientists at the Carnegie Institute in Baltimore were studying how the genes expressed themselves. It was the beginning of an era yielding a flood of information about this specific and odd form of DNA—the only DNA found outside a cell's nucleus.

It was a molecule tailor-made to Brown's key interest: the relationship between distantly related organisms. How did animals and humans relate? Maybe you could chop up their DNA and compare the fragments. His mind was still on the sort of questions that had him going as a kid. "I had a cell line from green monkeys and made some mitochondrial DNA from those and compared it with humans," says Brown. "It was not identical at all. It was at the outer edge of similarity by this particular assay. They were greater than 20 percent different. That was a mind-blower because it was absolutely contrary to what I and everyone else I had spoken with thought would be the most likely outcome."

These were the early halcyon days, and Brown soon moved to the forefront of comparative research. In 1974, he and Vinograd wrote a paper for the *Proceedings of the National Academy of Sciences* describing how a technique of mapping animal mitochondrial DNA was used to determine the size and order of DNA fragments drawn from mice, monkeys, and humans. It wasn't something to take to the beach with you. There were all kinds of strange language—"φx174 RF DNA," for example, was an adjective—and when plain English was used it still had an almost extraterrestrial ring. The punch line could have been spoken by the slow, metallic voice of a computer. ("The development of six cleavage maps is described in this communication. It is anticipated that these maps will serve as primary maps for subsequent mapping studies.")

Like the work of Wilson and Sarich, the mitochondrial DNA reduced itself, for now, to a molecular comparison of different animals. But it was much higher-tech. Instead of antibodies there were endonucleases. Instead of blood protein there were far smaller nucleotides.

A nucleotide is the basic building block of DNA. It is made of a sugar, a phosphate, and a base. The bases on one strand link with those of its partner. There are 16,569 nucleotide base-pairs (composing 37 genes) in the married (that is, paired) strands of it. Specific enzymes or "endonucleases" recognize a sequence of nucleotide base-pairs and chop the DNA there, causing fragments that are easier to study. Chop them up and put them under an electron microscope. An enzyme is like a key that fits only a specific lock. A single enzyme can settle upon the chain of 16,500 nucleotide base-pairs and break them into, say, eight fragments.

Once strands of mitochondrial DNA were broken into manageable parts, it was then possible to compare one organism's—or one person's—fragments to another's. When there were differences, these differences seemed to be caused by those telltale mutations. A mutation, in many cases, means only that one kind of nucleotide base-pair—there are basically four—has been replaced by another kind of base-pair or the sequence otherwise altered. It is easiest to think of the nucleotide types as A, C, G, and T, which stand for adenine, cytosine, guanine, and thymine. Viewed as a single strand, one person may have a sequence that looks like ACCT instead of ACGT. The "G" has been replaced by the second "C." It is a change in the ordering. Other mutations may be in the form of base-pairs that reverse or disappear. There may also be additions. And over the entire 16,500, whole sequences can be added or deleted.

Although the early workers looked not so much at the precise sequences as the gross, overall differences between two different DNA fragments (which indicated mutations and could be expressed as percentage differences), they nonetheless were able—people like Wes Brown—to connect one person or one population or one ethnic group to another by the overall patterning and similarities. Whether male or female, everyone can be related to someone else by connecting female lineages based on mitochondrial mutations that are shared between lineages. The longer two people have been separated from a common ancestor, the fewer patterns they would share. That is, the more mutations would separate them.

In mitochondrial DNA, coming as it does solely from the mother, with virtually no tumbling of the marbles, mutation is basically the only kind of change that can occur. Using enzymes to break the DNA molecule into fragments, the pieces are then placed in a synthetic gel and subjected to an electric current. This procedure, called electrophoresis, causes the nucleotide fragments to line up in order of size. The fragments are then separated by molecular weight, and dyed with radioactivity so they can be better seen. The result is a genetic "signature." Through comparison of these signatures, by mapping the fragments and estimating the mutations from them, it is possible to compare one animal with another—or one human with someone else—and tell just how closely they are related.

Wes Brown had a whole little arsenal of enzymes to chop DNA into discernible fragments. They went by esoteric names like *Eco*RI and *Hind*III. With them he was able to write the first paper on fragment patterns in monkeys, apes, and humans. Compared to the study of proteins, looking at DNA was like increasing the power of a microscope ten or a hundred times. It was the DNA and its sequences of nucleotides that determined, after all, what amino acids would be formed by the cell, and amino acids are what constitute a protein. There was even a slogan for it: "One gene, one protein."

While the genes in the mitochondrion are nowhere nearly as numerous as those in the nucleus, they are nonetheless a fairly exquisite signature of individuality. They also seemed—at least in the early days, to certain researchers—to be a damn good clock. They "ticked" faster—accumulated more mutations—than the genes in the nucleus. Brown discovered that mitochondrial DNA mutated five to ten times faster than nuclear DNA. Every few hundred thousand years it underwent significant change. That made it a particularly nice gauge of short-term evolution and the most precise clock available for telling when humans first diverged from a common ancestor, branched apart, and developed their own patterns of mutation.

If the mutations could be tracked back through time, their number would theoretically become less and less, until eventually there would be a point at which there had been one and only one type of mitochondrial DNA, and that would be the point of a common female ancestor.

Five years before his paper with Vinograd, Brown had learned of Sarich and Wilson during a symposium at Cal Tech. Wilson was doing the speaking. It was in the midst of the ruckus over albumin. And when Brown heard Wilson linking molecules to evolution, his ears pricked up like a rabbit's. "The contention that Vince and Allan made that human and gorilla and chimpanzee were only separated by five million years was really radical," he recalls. "There was a lot of vituperation about Sarich and Wilson, who I'd never heard of before. There were people in the audience frothing at the mouth, saying it was crazy. When I say people in the audience were frothing at the mouth, I mean, they really *were*. That was the summer of 1969. And that's when I started to follow the work coming out of the Wilson lab."

It would be a while before Wes Brown actually joined forces with the Berkeley group. When Brown was looking for a place for postdoctoral studies, he visited Wilson's laboratory but saw that it was "low-tech." It was still geared for proteins. They needed someone to set everything up. They were still shooting up bunnies to make antibodies. There were a few water baths and a spectrophotometer and low-speed, Maytag-like centrifuges, but that was about it. The place was not ready for DNA, and that turned Brown off; he was too busy building a basic foundation of expertise as a biologist to spend his time establishing a new laboratory.

Instead, Brown went across the bay to the University of California at San Francisco, where he picked up cloning technology and fell under the tutelage of a biochemist named Howard M. Goodman. They kept getting closer to the secrets of mitochondrial DNA. Day by day, month by month, the genetic material continued to reveal its orderings, functions, and regions. It was like going into a new city and getting to know it block by block. For lack of a better metaphor, consider the nucleotide base-pairs as individual houses, the genes as voting districts, and the entire DNA genome as the metropolitan area.

At the same time Brown was collecting his information in San Francisco, at Stanford University in nearby Palo Alto a man of about the same age, Douglas Wallace, was equally fascinated with mitochondrial DNA's peculiarities. Wallace had been involved in mitochondrial DNA since 1971, and he too had been asking fundamental questions, like how many copies of it were in each cell. It was Wallace and his Stanford group who would show that in humans it was definitely inherited only from the mother. Wallace worked with Luigi Luca Cavalli-Sforza, an aristocratic professor (now chairman) in the genetics department who had been one of the very first to apply high-tech genetics directly to human anthropology. In 1963, before Wilson and Sarich were even comparing proteins from apes, Cavalli-Sforza, then associated with the Universitá di Pavia in Italy, had begun using a computer to estimate the genetic distances between blood group systems from fifteen populations and build an evolutionary tree. Cavalli-Sforza had found, using this rather crude form of measurement, that Asian populations seemed genetically quite separate from Europeans and Africans.

Brown too was interested in human ancestry, and to investigate it he would need a lot more mitochondrial DNA. He realized there was a rich reservoir of such genetic material in human placentas—the nutrient-carrying organ that connects a fetus to its mother—and he began collecting such tissue from the maternity ward at the university's hospital. It wasn't always the most pleasant or neatest of tasks. Another name for placenta is afterbirth. "At one point I went up to the delivery room and the lady said, 'Your bag's in the refrigerator.' It was where they kept tissue specimens for biopsy. And there was a big bag there with my name on it, and I took it down to my lab and when I unwrapped it, there was a dead baby."

Most of the time, however, the placentas were collected less eventfully, and once collected they were ready for the enzymes. He ground up the placentas, extracted and purified their mitochondrial DNA, and chopped the DNA into those manageable pieces.

In 1979, Brown reported the electrophoretic analysis of genetic material from twenty-one humans of diverse racial background. The paper, entitled "Quantitation of Intrapopulation Variation by Restriction Endonuclease Analysis of Human Mitochondrial DNA," was filled with titillating little nuggets. Employing eight restriction enzymes to cleave fragments from the mitochondrial DNA of Caucasians, blacks, and Asians, Brown learned that this DNA differed on an average of 65 nucleotide base-pairs out of the 16,569, or one in every 250. It wasn't that much of a difference, and since mutations accumulate with time, it implied that the common origins of these people had not been very long ago. It also raised the possibility that all these people came from a population that was small in size—a population that had long ago experienced what population geneticists call a "bottleneck."

The analysis also confirmed the fast pace of mitochondrial DNA's evolution.

This type of DNA, concluded Brown, "will thus be especially valuable for studies of genetic variability within and among populations and among closely related species."

If the paper was filled with potential, it was all but ignored by an anthropological community that was unfamiliar with the confusing jargon of biochemists and was still chasing lustily after fossils. The 1970s had been a banner decade for bones; finally, it looked like all the missing links were falling into place. Never before had so much dramatic fossil material been unearthed in such a short time span. Who needed the gobbledygook from biochemistry?

Much of the drama centered on the Leakey family in Kenya. Although the family head, Louis, was old and ailing and would not live much into the decade, his wife Mary was still active in the field, and his son Richard—galloping on camels to ancient digs—was finding extraordinary fossils at a pace that nearly overnight put him on a par with his father. Richard found not just bits of bones belonging to a man-ape or ape-man, but whole populations of them. Working near what is now called Lake Turkana, Leakey found hundreds of teeth, jaws, limb bones, and skulls— tallying to perhaps 100 discrete specimens.

In 1972, Bernard Ngeneo, a member of the Leakey research

team, discovered a complete cranial vault at Turkana plus fragments of the facial skeleton. It belonged to a humanoid that came just before *erectus,* and it was about 1.8 million years old. Announced in *Nature* the following year, the fossil soon became famous as "skull 1470" (its number in the vault room of Kenya's national museum).

Leakey's parents had found a similar specimen, but nothing this complete. The skull was reconstructed to much of its original wholeness. It was dark and mysterious and wide-eyed. While the brain, under 800 cubic centimeters in size, was at the very bottom end of the range accepted as the genus *Homo* (modern human brains are about 1,350 cubic centimeters), it was distinctly larger and more "human" than specimens like the vaunted Taung child. A big find indeed, the fossil made headlines around the world and the cover of *Time* magazine.

Not long after the discovery of skull 1470, another hominid was found that created an even bigger sensation. The discovery was made the same year—1974—that Wes Brown was starting to publish his first substantial results from dissecting the mitochondrial genome. This time it was not the Leakeys but an American named Donald Johanson. And it wasn't near the Leakeys' favorite haunts in Kenya and Tanzania but to the north, in parched, famine-prone Ethiopia.

On November 30 of 1974 Johanson, camped near a muddy river at a place called Hadar, found the oldest, most complete skeleton of any man-ape ancestor to date. It was an *Australopithecus* like the Taung child, but it was supposedly 2 million years older. Even though an adult, the hominid had stood only 3 feet 8 inches and weighed on the order of 60 pounds. Johanson collected 40 percent of the skeleton: jaw, skull fragments, two nearly complete arms, a spectacular set of ribs, vertebrae, about half the pelvis, and respectable pieces of both legs—an unprecedented number of bones for something of that antiquity. They nicknamed the fossil "Lucy" because it was female and because, at the beer celebration that night, the paleoanthropologists had played the Beatles song "Lucy in the Sky with Diamonds." More officially she was given a new classification of *Australopithecus afarensis,* which, along with

1470, was later to cause a controversy that continues bitterly to this day.

The Ethiopian discovery was far more electrifying than the image of laboratory technicians pondering little bands or blots of DNA. In the decade leading up to Indiana Jones, paleoanthropology may not have been a "hard" science, but it was certainly easier to make heroes out of fossil hunters than out of someone who bled rabbits for their antibodies or played with electrophoresis.

But what Brown was heading toward was not the old man-apes like what Johanson and Leakey were discovering but rather the origin of modern people—*Homo sapiens sapiens*—in the much more recent past. After plugging away for nearly three years in San Francisco, Brown crossed the bridge to Berkeley and decided to work with Wilson after all. Wilson was still champing at the bit to get the new DNA technology. He'd been writing about DNA's potential since 1975. Proteins were indirect and old-hat. DNA was where the action was. And Brown was finally in the mood to retool the Berkeley laboratory.

So the stage was set: Brown would show Wilson's workers how to go about analyzing DNA, and together, peering at a piece of material that was only a few hundred-millionths of an inch wide—a far cry from the size of a fossilized skull like 1470—they would come up, nonetheless, with an idea of where and how far back our DNA could be traced. In so doing they would come up with a new view of modern humanity.

Or at least Wilson would. Charismatic to begin with, Wilson, with the new DNA capacity, was just overflowing with adrenaline. True, he and Sarich had caused big waves declaring the point at which hominids had split off from the apes, but there was bigger potential in this stuff Brown was talking about, for if mankind could be traced back to a tiny population, maybe that tiny population was Adam and Eve!

What a joy, this DNA. What a special joy, this genetic stuff from the mitochondrion. It seemed so plain and straightforward. Where the nucleus was like sand mixed from any number of beaches, the mitochondrion was an organelle of dramatic and pleasant surprise, a bean-shaped maker of energy, suspended in heavenly protoplasm, mutations ticking away, containing precious secrets from uncharted centuries; the sand in this box from one beach and one beach only, the grains—the base-pairs—just sitting there in every living human waiting to be counted.

Except for the mutations, mitochondrial DNA was basically like it had been thousands of years ago—before automobiles, before gunpowder, before the Pyramids!

And if proteins mutated at a fairly constant rate, so must the DNA that made them.

In mitochondrial DNA, there was clearly the prospect of a new and better timepiece.

So Wes Brown headed for the freeways and the earthy, even seedy, little offshoot of San Francisco and Oakland known as Berkeley. There were still remnants of the sixties, and would be for years to come. If hippies were being replaced by Yuppies, if ponytails were out and Jim McMahon glasses were in, if Abbie Hoffman had given way to Michael Milken, who had already graduated from here, there remained the musty, misty aura of counterculture. Years later there were still guys playing street harmonicas outside the biochemistry building, and a Xerox place called the Krishna Copy Center.

Brown arrived at the Wilson lab in July of 1978, still awaiting publication of his work on the placentas. He and Wilson wrote a grant proposal to the National Science Foundation (NSF) and it "sold like hotcakes." Wilson had a Midas touch. That year he was even able to get $52,165 from the National Institutes of Health (NIH). When the NSF money came, Brown bought three ultracentrifuges and began training people in his technique. One of them was a very bright student of Sarich's—Becky Cann.

Soon papers were flying from the Wilson lab like streamers of confetti. That was another talent of Wilson's: getting stuff into print. He knew how to catch an editor's eye. And now that *Ramapithecus* was being tossed out—and his 5-million-year time for the ape-human split accepted even by formerly reluctant paleoanthropologists—Wilson experienced a windfall of credibility. He and Sarich had stuck to their guns and now it looked like they had been right: man had diverged from chimpanzees 3 to 7 million years ago, giving the median age of 5. Although the low end of the range didn't fit very well with "Lucy," basically everything flowed into place and a lot of people believed what Wilson now had to say, which inspired him to keep saying something else. And Wilson—albeit from his hidden little alcove—would certainly have much more to say. Apes and hominids were one thing. Now he was on the trail of our mitochondrial mother.

Brown wasn't entirely comfortable with the way Wilson threw out assumptions and speculation. There was still one helluva lot of comparative work to do. He was soon up to eleven enzymes and counting. In short order he was finding out more about mitochondrial DNA's fast evolutionary rate, and learning about tricky little back mutations. The papers were finding their way into publications such as the *Journal of Molecular Evolution* and *Genetics*. Of particular interest was a paper Brown issued in June of 1980. In it he discussed some additional analyses of the mitochondrial DNA from his hospital samples, which had been preserved in a refrigerator. The samples were from ten white American women, two Filipinos, four Chinese mothers, an Egyptian, and four blacks from the United States. This time eighteen enzymes were employed to chop the sequences into understandable patterns, and what Brown found was that the mitochondrial DNA appeared to have changed at a rate of around 1 percent every million years. He and Wilson had determined this the year before by comparisons of human sequence divergences with chimpanzees, among other mammals. Using the 5-million time of ape-human divergence as a scale, as well as other divergence times based on the detested fossils, they had been able to come up with a scale onto which the percentages of mutational change could be superimposed.

They had calibrated a new clock: 1 to 2 percent change every million years. As was the case with albumin, you could then project back in terms of real time. Decide when one population had split from another, check out how different they had become over that period of time, and you had a rate. Again Brown mentioned the possibility of a recent population bottleneck: at some point the population stock leading to modern humans appeared to have been very small, either because it had been reduced by famine or some other drastic cause, or because it was just a small population to begin with.

Whatever the case, the origin of modern man wasn't a million years ago (as some paleoanthropologists, especially those who followed a French school of thought, were ready to believe); in fact, the origins seemed at least three or so times more "recent." Many of the mutations in the DNA Brown analyzed were shared by two or more of the samples, for which Brown could only rationalize a couple possible reasons. Either these genetic alterations had been acquired by coincidence—mutating in parallel with each other, which seemed much too coincidental—or they were inherited from a single common ancestor.

The important point was that *Homo sapiens* appeared to have originated from a limited group of people and to have done so in the *fathomable* past. That's correct: there was now a way, it seemed, to fathom modern man's specific origins. Genes in those odd circular strands called mitochondrial DNA had survived the millennia and basically had remained intact, descending through the centuries like a pocketwatch hidden in the anterooms of the cell, inherited and reinherited with little change except the tiny mutations, ticking slowly but surely, a time capsule in each and every living human, granting an unprecedented glimpse of the past.

At least that's the way someone who wasn't overly conservative could look at it. Away from the tumbling turmoil of the nucleus, the DNA was very much like it had been a quarter of a million years ago. With the rate of mutations, Brown could start extrapolating out specific periods of time, as Sarich and Wilson had done with gorillas. "At this rate the amount of sequence heterogeneity observed, .18 percent, could have been generated from a single mating pair that existed $180-360 \times 10^3$ years ago," said Brown, "*suggesting the possibility that present-day humans evolved from a small mitochondrially monomorphic population that existed at that time*" (emphasis added).

And in the end, because it was inherited solely from the mother, it had to go back to some single, solitary woman who lived 180,000 to 360,000 years ago, the mitochondrial mother of every living person.

6

ALTHOUGH HE continued to publish with Wilson for a few years, Wes Brown left Berkeley during the summer of 1980 and headed for the University of Michigan to start his own lab. As far as we're concerned, it was an abrupt disappearance. His voice would seldom be heard again. "At that point it seemed to me that I'd said about as much on the subject as ought to be said," says Brown. "I'd done my thing. I wasn't interested in grinding up another 150 placentas."

He also was not on the most congenial terms with Wilson, who was forever coming up with what Brown thought were pretty wild ideas.

Brown turned his analysis of human mitochondrial DNA over to the student he had helped train, Rebecca Cann. She was fascinated with the problems and challenges raised by the new research. Brown had said more data was needed, and Cann agreed to enlarge the sampling. There was that hypothetical mother back there, awaiting discovery, and that's what she was interested in, the origin of *Homo sapiens sapiens*. It perplexed her that more people were not studying the specific emergence of modern humans as opposed to the obsession with much older creatures like *Australopithecus*, which was at least half animal and was not an immediate ancestor.

Becky Cann had a laid-back air about her. Short of stature, she was the type who wore a floppy smock and let her brownish-blond hair flow where it wanted to flow, and she sat in a wicker chair. She had a demure, earth-mother countenance, and a ready, wrinkling smile—but with a cutting edge. She had the habit, as many geneticists and anthropologists do, of pasting cartoons and odd humorous headlines on the doors of her laboratory. Mainly they were from ridiculous supermarket tabloids. "Chimp brain put in human," said one, while another headline announced that Adam and Eve had been found, of all places, in Asia.

A joke, pasted up there with a sense of humor that was congenial and wry and biting. There was also a balloonlike replica of an ape in her office and the poster of a platypus. This was at the University of Hawaii, where she had gone after *her* stint at Berkeley. In the backdrop were mountains where mists and dark primordial clouds clung constantly to the upland, yielding an after-noon sprinkle of rain.

Cann had come to Hawaii in January of 1986 in the midst of finishing the *Nature* paper. Off and on, including undergraduate studies, her association with Berkeley had spanned fifteen years. Now she was setting up her own shop—"scrounging, scrimping, begging" for equipment and funds. Already she had a $40,000 centrifuge to work with, and the DNA was in the refrigerator by the door—in the butter compartment.

Like Brown, she was extending interests she'd held since childhood. Cann had been born in eastern Iowa, and on the way to Grandma's you could find fish fossils in the low, craggy bluffs along the Mississippi. She was interested in all sorts of organisms, but it was the idea "that things were not always as they are now, that there were once inland seas," that especially intrigued Cann. When she was fourteen her father, a stockbroker and jack of several trades, moved the family to San Francisco, where a young, inquisitive Cann developed new curiosities, marveling at all the different types of people there were: Koreans, Filipinos, Chinese. "That's when I first realized," she says, in the curious language of a scientist, "that I was interested in humans."

Cann never bothered with the cheerleader squad, but she was elected president of her class, had that quick sense of humor, and went to the proms. She also liked to play with words, writing plays. It was a fairly normal social life, not overly bookwormish.

"I wasn't an intellectual nerd," she says. "I didn't run into those types until I got to Berkeley," which was in 1969, as an undergraduate taking biology. It was a great time to be a scientist, and an exciting time for biology particularly. It was the year America landed a man on the moon, and it was also the year scientists isolated a pure gene. A watershed time for science as a whole, which had reached a historic peak—and now seemed all but infallible. In the gene would be the secrets of life. It seemed overwhelmingly important. On everyone's lips was the magical acronym, "DNA."

There was still political excitement on campus, and walking there one day, on the way to a course, Cann encountered tear gas, but she wasn't much of an activist herself. She was more interested in lectures by Glynn Isaac, a paleoanthropologist, than in burning a flag. She took courses on prehistoric man as a treat. They held her over while she labored through a genetics major. Unfortunately, Isaac's lectures were suspended during the protests over Cambodia.

As a junior, Cann took one of Wilson's biochemistry courses, but they didn't get to know each other until later. Wilson seemed shy, less than Berkeley's most dynamic lecturer, and his class, like most classes at Berkeley, was huge and anonymous. After graduation Cann had married and worked for five years at a pharmaceutical firm that made insect repellent. She sneaked in some reading in the mailroom and continued attending Berkeley, enrolling in graduate school. Her adviser was Vincent Sarich, who still held court in the Wilson lab.

At first Cann found Sarich "bombastic" and intimidating. It wasn't until later that she would see his warmth. He liked starting arguments for the sake of starting arguments, and he was pretty sure he was brilliant. There was also his towering height— six foot six—which Cann, ever the analytical mind, calls his "interesting morphology."

Cann was chosen to be one of Sarich's students because she had written about methylation and the control of DNA sequences. Sarich was one of the few who knew what she was talking about, and he, in turn, needed her expertise. He wanted to look at more monkey proteins, and Becky knew techniques like electrophoresis. She worked with Vince on Old World primates—macaque monkeys—and then went to Stony Brook, New York, to learn a DNA technique known as "southern blotting." What they were always looking for were new genetic markers. They were soon to find a treasure of them.

When Rebecca returned to California, there was Brown, retooling the laboratory. New equipment and new faces were everywhere, gearing up to exploit mitochondrial DNA, and there were frogs and lizards and samples from gibbons. It didn't take much to convert her. She saw the mitochondrion as a wave of the future. "Nuclear genes usually didn't retain enough information, they changed too slowly, you had to look at a lot of DNA from a nuclear gene to make sense of what happened," she says. "But if you could work with a tool like *mitochondrial* DNA that experienced mutations and changed very quickly, the possibility was that there would be a lot of DNA differences to detect and you could make headway quicker."

In a broader sense mitochondrial DNA also held hope as the way to spur interest in the more recent origins of man. Lucy and skull 1470 were highly dramatic, but they weren't *people.* They were half-apes going back way before the advent of modern man. Yet that was where the action was. There were more students studying man-apes than there were actual fossils to study. It was crazy to Cann, and worse, there was a sharp political edge. You might not get to see certain relevant bones if your theory was different from that of a fossil hunter.

Mitochondrial DNA, on the other hand, was democratic in the sense that anyone who set up a lab for it could produce his or her own findings. "It was clear," she says, "that molecular technologies were going to break open that field."

Cann's fascination was the transformation of the thick-boned *erectus* into gracile *Homo sapiens*—not man-apes. "It's not a question," she says, "of looking at something that was half animal. Instead, [with *sapiens*], you might be able to imagine the same kinds of motives, the same kinds of fears, the same sort of thought processes in those people as culture became more important and family became more important. And the archeology was there, the sites were there, you could see people living in what were social groups."

People, not australopithecines.

She had a great optimism that she could help formulate a specific view of that closer past. The Wilson lab was intense and productive, filled with people studying molecular adaptations and systematics and all sorts of new things. The lab was also irreverent and renegade. "Allan had this real rebel streak in him, and he didn't necessarily want to know what an Ernst Mayr or a Dobzhansky had to say," remembers Cann. "Those guys were wrong and old fogies, and you shouldn't pay attention just because they had these immense reputations."

In the beginning she was an unfunded teaching assistant, but she had the lab at her disposal and that whole flock of fellow thinkers flitting amid the flasks. Everyone was fascinated by what Brown was tossing around: this concept of a population bottleneck (an idea spawned by the fact that humans seemed to have far less diversity between each other than did gorillas). While the average mammals differed by 1.5 percent in their mitochondrial DNA, humans differed from one another by perhaps 0.18 or 0.4 percent, a quarter of the average mammalian diversity, and less diversity raised the possibility of a small founding group. The smaller the group, the faster and more dramatic, perhaps, was anatomical change.

Although Brown would later withdraw much of his support for the notion, the idea that a very small number of founders established our current population—that there was a single area of the world where a band of people gave rise to us—was much too exciting to ignore, just the type of thing that sent Wilson's imagination flying. He mulled it over, came up with his own ideas, and gave the lab direction from his control tower behind the stage—the guru, reviewing papers and approaching the big conundrums and writing grants, all with the same earnestness and sense of mission that had propelled the albumin research into ultimate acceptance.

On the daily, nuts-and-bolts level, however, the human project became Cann's. She began tackling the issues that Brown's work had so brilliantly raised. Her main problem was getting additional DNA samples. She needed more placentas since all that was currently on hand was what Brown had given her. "I spent a long time trying to get donors, trying to get the cooperation of labor and delivery rooms," Cann says.

And the lab had to keep moving. There was competition just an hour and a half down the freeway, at Stanford, where Doug Wallace was gathering results proving a maternal mode of inheritance and was just a year or so away from publishing a paper about blood samples culled from 235 individuals from around the world. Even though placentas were something routinely thrown away, a hospital wouldn't release one to her without the specific permission of the mother. Cann wrote to expectant mothers, cajoled them, and spread the word through acquaintances. She also convinced a man in western Australia to get Aborigine samples. It wasn't easy getting placentas from people in remote outbacks. Aborigines were up in arms about geneticists who drew blood from their secretive societies or anthropologists who studied skulls and then ran off to say how primitive these people were.

There was also the spiritual side. "In New Guinea, in many places in Indonesia—Melanesia, Polynesia—blood samples and placentas would be considered immense sources of spiritual power, *manna,* and you wouldn't talk about it," says Cann. "It's secret, it's buried. When ancient Hawaiian chiefs died, their bones were hidden. Various scientists have gone to take blood in various parts of the world and sometimes blood sitting on a loading dock gets stolen. A lot of people working in remote areas are constantly telling horror stories about losing their samples. We assume that they're lost in transit. *Not always.* So it's a sensitive issue as to how you get the samples. In my case, the donors were pregnant women who were coming into nursing stations."

The majority of samples, however, were from the United States. One placenta came from the wife of a lab worker. But even back at home it wasn't all that easy. It took more than two years to get a decent sample size. Most people looked askance at a young scientist pleading for pieces of afterbirth. "It took them a while," says Cann, "to figure out that we weren't lunatics. At first most placental samples came from women who were friends of friends. People would go to parties all over Berkeley and they would scout

out pregnant women. Friends recruited for me. It was a grass-roots effort. Word spread through Lamaze classes. It took a while to work into labor and delivery rooms. Labor and delivery people are busy, and the last thing they want is another person telling them to save something. A lot of samples came from the university hospital in San Francisco. I was calling every morning to find out if there were new births, and going over to talk to the mothers to see if they were interested. I was just exquisitely slowly waiting for these babies to be born."

Although the mothers whose placentas she sought ended up being very cooperative, at first she got the okays from only about half of them. The cool and analytical mind of a biologist was not always in sync with an expectant mother or one who had just gone through the tender and exhausting experience of giving birth; scientists often had that unnerving way of expressing themselves. "Talking to new mothers was really an interesting experience because this was a point where I didn't have children and I wasn't interested in reproducing myself," says Cann.

"I didn't know many people in my socioacademic setting who were producing babies and I had a lot of prejudices against people who were."

It wasn't exactly the picture of warmth, but eventually she got her placentas. They were added to the twenty or so from Brown. By the time Cann completed her thesis, along with a paper that appeared during 1984 in the journal *Genetics*, there were a total 112 samples of tissue. Most, again, were from Americans, and that meant a varied mix. There were samples labeled Swedish/English, English/Irish, and German/English, and also ones ascribed to women of Nigerian, Mexican, Vietnamese, Chinese, South African, H'mong, and American Indian descent. Since Cann had not been able to get to Africa itself, they decided to use American blacks as representatives of Africa. This was a decision that was controversial in the extreme. Blacks in America had mixed with Caucasians through the last couple centuries, and many had white blood in their veins.

The work wasn't for the weak of stomach, even if there were no more dead babies in the pickup bags. "Becky was constantly coming in with these bloody placentas and occasionally freezers would break down and you would have blood leaking out of a freezer onto a floor," former lab worker Steven Carr remembers. "On one occasion she lost her key or for some reason couldn't get into the lab, so she stashed her placenta into some place you would not have expected to see a placenta and someone got into that freezer and was rather horrified to see this mass of bloody tissue."

Carr glued a Styrofoam container to the door as another inside joke. "Deposit placentas here after 5 P.M.," it said.

The lab was an odd mix of geneticists, biochemists, and Becky Cann, who was not only a geneticist but also an anthropologist. Since they were working with DNA, it seemed easiest to call them molecular biologists or geneticists. Another lab worker, Cann's co-author on the *Nature* paper, Mark Stoneking, had been collecting samples from the remote highlands of New Guinea. Actually, he was the one arranging the most pickups at the airport. Stoneking was six years younger than Rebecca, a quiet man, slight of build, with piercing, gelid eyes. His background was similar to Cann's. He had an anthropology degree from the University of Oregon and a master's in genetics from Penn State University. Through the help of a contact at the Papua New Guinea Institute of Medical Research in Goroka, K. Bhatia, who received, collected, and shipped the placentas, Stoneking added a couple dozen more samples to the collection, the last shipment arriving during 1984. The image of someone in the remote hillsides sticking bits of placenta in tubes tickled Becky. That *was* getting closer to the "Temple of Doom." The placentas were frozen, thawed when they were ready for analysis, chopped for samples, treated in a petri dish, subjected to scalpel and razor, then dumped into a blender until the cells burst, turning the concoction into something the disconcerting color and consistency of a strawberry daiquiri.

"We'd have these Bloody Mary or strawberry daiquiri parties, and people weren't so sure what they were looking at," says Cann. The basic idea was to crack open the cells and separate DNA. They'd take the "daiquiri" mix, spin it in a low-speed centrifuge at less than 3,000 revolutions a minute to pellet out the nuclei and other junk, add solvent precipitate at a strategic moment, and spin the homogenized placental tissue at a faster speed to get rid of platelets, bits of membrane, and other cellular debris. That was the "crude prep," and it still needed more cleaning. There was additional spinning and other rituals of biochemistry. Eventually they'd end up with a concoction that looked more like apple juice than a strawberry cocktail. That was the cell's DNA. It was drawn into tubes and subjected to more spinning, this time in an ultra-centrifuge that revolves at 36,000 times or more a second, creating a force about 300,000 times that of normal gravity. At the end of *that* spinning, Cann or Stoneking would turn on an ultraviolet light and readily distinguish the DNA by a pink hue caused by fluorescent dye. The mitochondrial DNA was just below the nuclear DNA. It was drawn out with a pipette, cleansed of its salts and dye, mixed with an enzyme that chopped it up according to nucleotide sequences, labeled with radioactive material that defined end points, and put into a box of gel that looked like clear Jello and was subjected to an electrical current. The current sorted out the fragments and lined them up according to size. The gel was then exposed to X-ray film that took pictures of the fragments, which looked like bands with various shades of darkness, and these pieces of DNA were painstakingly mapped, compared, and otherwise analyzed.

Most important was the intricate and astonishing process of calculating mutations by the size of certain fragments. By this time Cann had an advantage over Brown's previous studies in that a group from Cambridge had published the complete sequence of a human mitochondrial DNA, allowing ready comparisons. Also, Cann was using a mapping method that had higher resolution. Where Brown had broken his samples into 50 fragments, Cann could examine 370. They were called "restriction sites."

Another way to look for mutations was a method known as sequencing, in which the base-pairs were actually tallied individually, but that was a tedious and long, drawn-out process. Cann preferred to get a bigger picture with the fragments and estimate the changes and differences. At any rate, both were ways of seeing what the eye couldn't possibly see. What Cann and now Stoneking were using for their calculations was a better array of enzymes, fixing more landmarks along the ostensibly featureless but really dauntingly complicated terrain of DNA. With better enzymes or sequencing techniques, more minute comparisons could be made. And the better the method, the surer they were of mutations detected in this chemical soup of genetic punchtape. These tiny mutations, which cause one person's DNA to sometimes appear slightly different from another's (and certainly different from a gorilla's), are, of course, the genetic markers Cann had been looking for.

It has been speculated that many mutations occur as accidents when the DNA is replicating itself, a mistake in copying, perhaps due to an interfering agent that may knock atoms off the DNA molecule, changing the chemical composition of the base. During the course of a million years, mitochondrial DNA may accumulate something on the order of 20 mutations for every 1,000 bases. Though these mistakes may be a random mispairing of base-pairs during replication—a failure of DNA to exactly copy itself—they may also be the result of agents in the environment. Candidates for causing such alterations include naturally occurring radioactive compounds, environmental chemicals known as mutagens, ultraviolet radiation from a hot sun, or the highly penetrating and ubiquitous cosmic rays. Others have gone so far as to speculate that agents in mold eaten by humans can alter DNA, and so can the chemical composition of burned meat. In the mitochondrion most such mutations are meaningless to the cell, constituting little more than genetic noise, but they are far from meaningless to the evolutionist who believes that such mutations occur regularly through time and are the yardstick for how closely one organism is related to another—and, again, how long it has been since the two separated from each other.

It took up to two weeks to get pure mitochondrial DNA each time they went through the process. There had been frustrations and lost hours. In one case a whole month's worth of purifying went for naught. For two months in another case the film was blank and they couldn't figure out what was wrong and finally Cann realized she'd forgotten an ingredient that functioned as a stabilizer. Every once in a while a buffer would go off or a gel wouldn't work or a tag would dissolve so she couldn't tell who the donors were. DNA is small stuff, and they had to go through much of the process blindly, moving through perhaps twenty steps without actually being able to see what they were producing.

Trying to map restriction site changes, seeing where the mutations were, consumed huge amounts of time and tried even a scientist's formidable patience.

But the little thrills, when they came, more than offset the tedium. Sending your first base sequence through an autoradiograph was a real high. And the first results indicated that blacks, whites, Asians, and Australians were, on the whole, very similar to each other, differing by about one in every 400 bases.

The labor involved in assessing such minutiae often ran through the night. There were questions about things called transposons and the mobility of sequences in the mitochondrion and all kinds of other nuances that escaped anyone without a centrifuge and autoradiograph. One mystery had to do with a section of DNA that seemed to change in length like an accordion, expanding and contracting. It was a little mystery novel, the mitochondrial DNA. There were all kinds of other subplots going. And one of them led to a significant mistake.

"When we first started seeing the Australian Aborigines there were some unique—what I *thought* were unique—base changes or restriction patterns that we'd never seen before," recalls Cann. "And this is an area of the world that people were saying was an area of old occupation and we all knew that there were bizarre fossils. The question was always, were there any relic DNA lineages that would be present if we started sampling on a worldwide basis. Where were we going to see genes that last evolved in a species that was not *Homo sapiens?* And so we were constantly on the lookout for what was fairly anomalous. As I began mapping, I started to see these really, really strange patterns that corresponded to what we now know are length changes: insertions and deletions."

There were other curious traits as well, including reports of rare alleles (alternative states of a gene originally produced by mutation) in the nuclear DNA. But before they were entirely sure why some of the Australian samples seemed so odd, Allan Wilson had taken the preliminary data and trooped halfway around the world, interpreting the results as possible evidence that the origin of modern man was a place just about no one expected: Australia.

Wilson thought maybe a confusing bunch of mutations was the signature of an ancient ancestor. Since certain of the Australian samples showed characteristics that weren't observed in other populations, there was the possibility that such genetic traits were the remnants of *erectus* himself. Late the night of June 21, 1981, in the bar of King's College in Cambridge, he told a science writer, Jeremy Cherfas, who taught zoology at Oxford, that although it seemed "far out" and most people would consider what he had to say "so unreasonable that it would prove how crazy we are," there was a possibility that "while *erectus* was muddling along in the rest of the world a few *erectus* had got to Australia and did something dramatically different—maybe not with stone tools— and that that's where *sapiens* evolved and then got back to the rest of the world." As Cherfas later wrote, this implied that Homo *sapiens* "evolved, free from competition, out of a small band of Homo *erectus* that, 400,000 years ago, somehow made its way across the Timor straits and into Australia."

Wilson came out of the woodwork when there were deep thoughts to think, a paper to write, or a grand hypothesis to explain. He was a quick thinker. A lab worker could knock on his door with some new result and Wilson, who no longer did bench-work himself, would have an interpretation of the data before the worker did.

But this time, six years before the mitochondrial DNA was ready for presentation in *Nature,* he had picked up the ball and was running toward the wrong end zone. *Australia!* It was actually just a miscount or misinterpretation of DNA variances. Australia was *not* our ancient homeland. It's not where Eve was.

Forgiven or forgotten by the scientific media, which was soon to telegraph his sub-Saharan claims, the Australia gaffe did not sit well with Brown, who had been reduced to watching from the sidelines as Wilson, Cann, and Stoneking began expounding upon the results Brown had initiated.

Wilson was an excellent biochemist, thought Brown, a man of many ideas, but some of those ideas were "off the wall" and Wilson, despite his reclusive behavior, seemed secretly to like publicity. "Allan went on this trip about the Australian samples and I guess started talking about how aboriginal DNA was different than other mitochondrial DNA and may mean that the Australian aboriginals were representatives of *Homo erectus* or something like that," says Brown. "What turned out to be the case after they did the proper science on it was that there was simply a deletion or addition of nucleotides that made things look much more different than they were."

Embarrassed but undaunted, Wilson stayed hot on the DNA trail. And soon indeed he would have more theories to propound, and a discovery, with Cann and Stoneking, that was in some ways more arresting than skull 1470 or Lucy.

THE FIRST INDICATIONS that a group of molecular biologists at Berkeley were up to something unusual came not in splashy tabloid font but with provocative discourse in the technical journals. It was insider's talk, a discussion of mutational rates, a chart of site variances for protein-coding regions, and a smidgen of self-promotion. ("Ours," said one paper, "is the most comprehensive study available for complete mitochondrial genomes that are extremely closely related.") Whatever the fossil hunters might think, it was Wilson's contention that "the molecules of life are now the chief source of new insights into the nature of the evolutionary process."

Although Brown was still a by-line with Wilson, Cann was taking his place just as he had earlier assumed the spot once held by Sarich. In 1982 there had been a preliminary report on the work in a densely academic book issued by New York medical publisher Alan R. Liss, but it was just that—a preliminary report—and ran to only nine pages. The following year Cann and Wilson published a paper on length mutations in the journal *Genetics.* In the 1984 paper, which included Brown's by-line, they suggested that "the pattern of sequence divergence inferred from our comparative study of human mtDNA may have several implications for understanding the forces that drive evolutionary change in mtDNA." It still appeared, they pointed out, that mitochondrial DNA did not get jumbled up by sexual recombination.

89

Little delights and surprises seemed always present in that minuscule cellular furnace—that microscopic sausage of an organelle—known as the mitochondrion. In some microphotographs, mitochondria looked less like sausages (the favorite description of textbooks) than they did amorphous little blobs of rain squiggling down a windshield. Either that or the deformed potatoes. And no real flame inside, but rather a place of oxidative reactions. A maze of microscopic corridors is what it was: infoldings around the matrix of enzymes and lipids and those stringlike loops of DNA carrying a Morse code from otherwise unapproachable epochs.

The molecular evolutionists, delving each month and each year deeper into the idiosyncrasies of mitochondrial DNA, spoke as if everyone knew the intimate features of genetic chemistry. It was maddening to wade through whole sentences that spoke only in superscript or acronyms. H-strands and D-loops and the genes. An urf6. Or an ND4. Meanwhile that DNA map from Cambridge was not the sort of map—not with whole pages of tiny letters and band schematics—that you would find in Rand McNally. Published in *Nature*, the paper delivered to scientists like Cann an invaluable source of reference, granting them a base of comparison for their own bits and pieces of protein and tRNA genes. What it came down to again was that the circular DNA strands divided into genes, and each of those genes was composed with hundreds of nucleotide base-pairs, so that a fraction of one might look something like "TGACCCTTGGCCARAATATGA."

But occasionally there were morsels of information that were more understandable if for no other reason than because they could be translated into dollar signs. For instance, the small print of acknowledgments at the end of some papers explained that part of the research had been funded by the National Science Foundation, which had given Wilson a princely $387,996 for a grant acknowledged in the 1984 paper as "DEB81-12412" (its bureaucratic code) and another $394,913 for a grant alluded to in a paper co-authored with Brown (DEB78-0284). The research also had been funded in part by the National Institutes of Health and the Foundation for Research into the Origin of Man—originally a

Leakey fund-raiser that supported Cann's very doctoral dissertation.

The taxpayer-funded NSF, a division of the federal government's executive branch, was also supporting other evolutionists like Morris Goodman and had supported Wilson's earlier gorilla work. The expenditures were difficult to fully reckon but it seemed fair to say that, in combination with grants from those other organizations, including at least one other federal agency and naturally the state of California (Berkeley is state-operated), the search for Eve, all told, including the primate studies and protein research, was starting to approach—or even exceed—the million-dollar mark.

That wasn't so much in the age of Donald Trump, but it wasn't sawdust either. To Wilson, however, the money was a means to a more important end. He drove the lab managers crazy. If there had been a time after the protein work that grants had been hard to come by, which made him a bit bitter, he wasn't in bad shape at all anymore and he wasn't the type to dwell on nickels and dimes. The important thing was to "do science." And if Wilson was a reclusive figure, carefully guarding his privacy and (when he grew his white locks) looking indeed the guru, he wasn't so shy about approaching agencies for public funds. Without them, the "science" would have been vastly more difficult to "do"—if not outright impossible.

But for now there wasn't much to worry about. They were on a roll. This was exciting science. It seemed like everyone was picking up the mitochondrial DNA ball. By this time analyses were being conducted on all sorts of animals, from fruitflies and tree frogs to reef fish and blackbirds.

It was so easily purified, this type of genetic material, and it seemed to be maintaining its unique properties. For one thing, it still looked certain that mitochondrial DNA was inherited solely from maternal lines. As Wilson, Cann, and Stoneking pointed out in a 1985 paper read at a Population Genetics Group meeting in Manchester (and published that same year in the *Biological Journal of the Linnean Society,* with the names of eight other researchers, so as not to have anyone feel left out), backcrossing tests with a pair of moth species had shown that after ninety-one generations, no paternal contribution was in evidence. Though they saw short-comings in such tests, all in all the indications were encouraging to Wilson and his earnest group. Meanwhile, other researchers like geneticist John Avise at the University of Georgia pointed out that the progeny of crosses between a male horse and a female donkey exhibited the mitochondrial DNA pattern only of the donkey.

Two years before, Wilson had written a paper with two other Berkeley biologists, Stephen D. Ferris and Richard D. Sage, presenting evidence that common laboratory strains of inbred mice were all descended from a single female. In the 1985 paper the authors mentioned that the mean rate of divergence—the pace at which nucleotide substitutions had accumulated during the course of evolution, averaged over the whole mitochondrial DNA molecule—appeared to be about 2 percent per million years in the gene pool of rodents, gallinaceous birds, geese, salmonid fishes, frogs, and the Hawaiian fruitfly. The same rate also seemed to hold true for mammals such as primates and the rhinoceros. Nucleotide substitutions, where one type of base was replaced by another, were, of course, a form of mutation.

Assuming that rate of 2 percent per million years, there was the ability, Wilson, Cann, Stoneking, and their crowd of co-authors pointed out, that "one can build temporal frameworks relating mitochondrial DNAs found within species." They drew an example of a tree and time scale relating types of mitochondrial DNA found in ten common chimpanzees and three pygmy chimpanzees. It showed, they said, that the mitochondrial DNAs *within* each species were more closely related than to the mitochondrial DNAs in the other species and also, more importantly, that *all* of these mitochondrial DNAs traced back to one female that lived perhaps 1.9 million years ago. That was before the pygmy and common chimps diverged from each other. The actual split into species occurred, according to further calculations, about 1.3 million years ago, in much the same way that the line leading to humans had split from chimps and apes about 4 million years before that. So the picture you had a million years ago was of chimps going one way, splitting into species, and humans in the form of *erectus* heading in a completely separate direction and ready to do some splitting of their own. Apes splitting into different kinds of apes and humans evolving into higher forms.

One peculiar thing about humans, the scientists said in that 1985 paper, was an "anomalously low" variability in their mitochondrial DNA. That implied that *Homo sapiens* were younger as a species than chimpanzees, since mutations were thought to be a product of time, and it also raised that possibility of a bottleneck in population size back many centuries ago. Apparent bottlenecks had been observed in animals like mice, and such small founding populations were the result, perhaps, of the last Ice Age, which kept animals away when ice sheets dominated the landscape and let them trickle back—at first in small numbers—when the ice melted. There was also, as always, the possibility of human influence. When the first farming families moved into Japan and Europe within the past 5,000 years, they brought mice with them. From a small population size—a bottleneck—the critters then blossomed forth into an expanding and permanent population.

"The whole point is that by tracing backward through time using techniques we have, it must be that eventually all the variation in a population would trace back to a single common ancestor," says Stoneking. "That's just basic evolutionary principles, that there is just a single origin for life, followed by modification and descent."

But it wasn't quite so clear-cut. In the laboratory itself there were differences of opinion. For the longest time Wilson, Cann, and Sarich (who served as a consultant) hadn't been able to decide, in terms of a time scale, just what they were looking at. Part of the problem was that computer programs did not exist that could handle the entire data set at once and allow them to construct a phylogenetic tree. Many of the computer programs were for proteins, but the matrix for DNA was larger, more complex. Though there were clear indications to Cann of an African origin back in 1982, with Africans more genetically diverse than younger populations, they couldn't put the entire data set into a tree-building program. And that's what they needed to confirm an African origin: a good old tree of evolution.

There was no internal calibration point to go by; they had to come up with their own. They couldn't automatically go to the

chimpanzee and use it as a comparison, charting time since apes
and humans had diverged and calibrating a clock that way, because there were no fully detailed and direct sequences of DNA
from chimps to align with humans. As for internal calibration,
that could come from comparing humans in certain isolated populations with each other. All you had to know was how long those
humans had been isolated.

It was reminiscent of the seminal problems Wilson had
approached during the albumin research. If he had stopped doing
daily benchwork, busy with visiting professors or writing new
papers, he was still fully in command of his large lab, behind the
scenes as the wizard of this Oz, orchestrating and questioning and
tossing off possibilities or theories like a Fourth of July sparkler.
"A lot of Allan's technique," says a former lab worker, "was to
throw out an outrageous idea and say, 'If you think this idea is
outrageous, prove me wrong.' "

Whatever the merits of his penchant for scientific slogans,
Wilson was precisely what most scientists were not: both cerebral
and entertaining. Those who criticized him—and there were at
least as many critics as there were followers—still acknowledged
his work as innovative and alluring. One day he would be able to
list on his résumé a Guggenheim Fellowship (in Israel and Kenya),
another one at Harvard, membership in the American Academy
of Arts and Sciences, and election to the Royal Society of London.
Even his bitterest foes conceded that he was a scientist of impressive capability, and the MacArthur Foundation certainly thought
so. In 1986 the Chicago organization awarded him a substantial
personal grant—$248,000—known as a "genius award."

It was like winning the lottery, and it also granted Wilson,
wounded by those early years of ridicule, a legitimacy beyond his
own inner circles, where he had always been held in high (if
sometimes nervous) regard. The awards are given each year to men
and women of outstanding and often unrenowned talent. That
placed Wilson alongside poets, mathematicians, and obscure com-
posers, as well as a roguish and controversial magician named The
Amazing Randi who made a name ridiculing psychics and was
striving very hard—abetted now by MacArthur dollars—to
become the next Houdini.

The mention of a normally anonymous scientist in the same breath as a stage magician was another triumph for the biochemistry department, which consciously or not had proven itself quite adroit at making splashes in the media. The chairman of the department, Bruce Ames, who shared the same floor as Wilson and communicated some of Wilson's papers, was well known as the inventor of a test for chemical mutagens and was becoming equally well known, especially among a most receptive local petrochemical industry, as the advocate of the interesting hypothesis that man-made carcinogens were no more harmful to the public than a peanut butter sandwich. Ames, the quintessential scientist with wire-rimmed bifocals, pale skin, rumpled suit, and tousled hair, was featured in publications like *Reader's Digest* for views that included not only the peanut butter assertion but also the headline-making claim that drinking a glass of the most polluted well water in Silicon Valley was a thousand times less of a cancer risk than a glass of beer or wine.

Or for that matter, one can rest assured, a strawberry daiquiri. It's what the lab workers drank when a paper was accepted for publication: if not those daiquiris, then the Bloody Marys, which seemed equally inappropriate, given the placentas in the freezer. But a party was a party, and there were more than enough reasons for celebrating. The Wilson lab, well funded now and with perhaps twenty workers, continued to crank out scientific documents like an assembly line. Cann had already published her dissertation on the first mitochondrial results back in 1982 and was now deeply immersed in the final stages of the *Nature* paper. Stoneking had brought in the New Guinea placentas, and their analysis—the genetic distance between these individuals based on the number of mutational or other differences—was basically complete: twenty-six samples mentioned in a 1986 paper and twenty-nine more in another paper published in the *Cold Spring Harbor Symposia on Quantitative Biology* that same year.

The New Guinea samples were crucial in calibrating the internal molecular clock, for it was one area where scientists figured they had a pretty good idea of when the place was settled. That colonization, it was suggested by fossil and archeological evidence, was at least 30,000 and perhaps 50,000 years ago. This means of determining a rate was in addition to the calculations based upon animal data, which included the primate-human divergence that Wilson and Sarich had determined to be 5 million years ago.

"Our calibrations of the rate of human mitochondrial DNA evolution are basically based on two lines of evidence," Stoneking explains. "One comes from interspecific comparisons— comparison of two different species whose time of divergence is either known or can be estimated from fossil or biogeographical evidence. Divide the amount of divergence by the time since they last shared a common ancestor and you come up with a rate.

"We also have tried to do a calibration specifically for humans. The way we've done that is to consider the amount of mitochondrial DNA sequence divergence within a population that is fairly isolated and remained in isolation since they colonized it, and under conditions such that migration back out to the original founding areas would have been minimal. Now there's no human population that's going to satisfy those requirements exactly, but the ones that come reasonably close are populations in the South Pacific—like the ones that colonized Australia and New Guinea— and populations in the Americas—native American Indians."

Using their own data on Australians and New Guineans, as well as data from that other major mitochondrial DNA researcher, Doug Wallace, who had been at Stanford looking at human platelets, they came up with the divergence rate of 2 to 4 percent per million years—which was about twice the rate Brown was determining.

It was also found that the nearest relative of a New Guinea cluster was either exclusively Asian or contained Asian mitochondrial DNA types. So Asia, long figured to be the staging area for colonization of Pacific islands, was confirmed as the source of New Guinea's mitochondrial DNA. There was no obvious association, however, between New Guinea types and Australian ones, implying, said Stoneking, Bhatia, and Wilson at the time (see Notes for another scenario), that Australia was not colonized via New Guinea nor New Guinea colonized via Australia, which was what many had thought, since New Guinea is the closest major landmass to Australia—and until 10,000 years or so ago, the two in fact had been part of the same landmass.

The picture beginning to take shape was of a single population that had spread around the Old World before any permanently distinct racial or ethnic groups had been formed. That meant there may have been a time when, on first sight, one might not have been able to distinguish a human living in what is now England from one living in what is now New Guinea. If the races had been differentiated long ago, all lineages within a race would trace back to a common mother who was also a member of that race, but that was not the case. "The common mother of all New Guineans was the common mother of all Australians and all Europeans," says Wilson. "She couldn't have been simultaneously a member of each of those geographical groups. So the process of the formation of human races is one that happened after the founding of the population by multiple mothers."

Such fascinating results became apparent as they devised what in the end would be the convoluted and horseshoe-shaped tree. By using the restriction sequence mapping, they gathered data that was then coded and entered into a computer. Each person was compared with every other person. Those with the fewest differences in their base sequences were grouped together in small branches, showing, in one case, a New Guinean next to a European—and far removed from other New Guineans. Literally thousands of branching diagrams could have been chosen, but using what is called the principle of parsimony, the tree was constructed on the basis of the simplest story that could be told. "We assumed," said Cann, "that the giant tree which connects all human mtDNA mutations by the fewest number of events is most likely the correct one for sorting humans into groups related through common female ancestry."

They placed the root at a point midway between the two most different lineages. That indicated Africa as the origin, because the two individuals with the most divergent ancestors both had an African ancestry, Cann explained. Furthermore, while there were Africans whose ancestry could be tracked without bumping into any non-African types, the descendants of the other areas all had at least one *African* ancestor. Africans, as Wilson had said, were the only ones on both sides of the tree, which had two main branches. One of the branches led solely to Africans, the other to all other groups and a scattering among those other groups of additional Africans (or at least American blacks).

Australia was no longer mentioned, except perhaps as a passing joke. Since 1982 there had been the growing—and once the tree was finalized, all but irresistible—indications that the place was Africa. The reason, says Cann, was because even with their small sampling of American blacks as representatives of an African lineage, "the level of diversity was so much deeper than the other groups." Indeed, Africans varied from one another as much as they collectively varied from any other racial group. They had accumulated a lot of mutations. Once again, more diversity meant a population had been around, in all probability, a longer time.

Still, there was a very troublesome point, and that was that they had not analyzed actual Africans but rather had used American blacks as a substitute. There hadn't been the money to get a good sampling directly from Africa, Cann says, and there were also what she describes as "political" problems gathering blood or placentas from that continent, a constant source of frustration and depression to her. At one point Cann had even wondered if the project was viable because of a problem in getting enough black samples. While American blacks, she figured, were likely to have an authentic African mitochondrial DNA lineage, there was still a big question, since American blacks had substantial quantities of not just Caucasian but also Indian genes in their nuclei, the result of admixture—sex between the races—once they arrived in the United States as slaves. Geneticists estimated that as much as 20 percent of black genes were white-derived, and in the case of Chicago, according to Cavalli-Sforza, perhaps as much as 50 percent of the black nuclear genome was of white descent. Though it might well not have had relevance to Cann's specific sampling (which upon later review showed little evidence of white mixture), a study was forthcoming indicating that there was perhaps 39 percent white input into the American black mitochondrial gene pool.

Yet Cann, Stoneking, and Wilson were not to be deterred. If American blacks were the only available sample for getting a fix on African lineage, American blacks it would be. There was a huge difference between the genes from the *nucleus,* where DNA came from both parents, and those in the maternally inherited and recombination-proof mitochondrial DNA. Black Americans, decided the scientists, would have African mitochondrial genes because sex between the races had been virtually one way and one way only: white males with black women. Since males contributed nothing to the mitochondrial DNA—and since, when there had been intermixing, it was usually a white man and black woman—the African lineage would have survived in the form of the female mitochondrial DNA coming from black women, no matter how much mixing there had been. Simply put, the African lineage would have survived intact.

This would later change from a frustration to something of a public relations nightmare. No one would accuse them of being racists (as Wilson had once accused the paleoanthropologists), and there was no indication whatsoever that they were.

But there was something about saying only white men and black women interacted when it came to sex between the races that had the ring of racial generalization. It also provoked ugly old images of slaveowners using young Africans sexually and leaving a child in the woodpile. As for the actual science of such an assumption, it was a small package of dynamite. Cann herself later described the assumption that only white men and black women had mated as "a seat-of-the-pants guess," adding: "And because most of the blacks in our sample were coming from the American population, they also represented the western African population primarily, and we didn't know much about East Africa or South Africa."

Using a different tack, Stoneking stonewalled the objections, characterizing them as "not being very rational. Given that mitochondrial DNA is maternally inherited and that the nature of the interactions between whites and blacks in this country would have involved for the most part, until recently, white males with black females, there's not going to be any contribution of Caucasian mitochondrial DNAs to the black American mitochondrial DNA gene pool. Furthermore, as we pointed out, there are a number of restriction site polymorphisms that appear in our black

American sample which are characteristic of native African populations, from work that Doug Wallace has done."

Of the 147 individuals in the formal study, fully 98 had been culled from American hospitals. There were 34 Asians (originating from China, Vietnam, Laos, the Philippines, Indonesia, and Tonga), 46 Caucasians (with roots in Europe, North Africa, and the Middle East), 21 aboriginal Australians, 26 aboriginal New Guineans, and the 20 "Africans," only 2 of whom, as mentioned before, were actually born in sub-Saharan Africa. The other 18 were American blacks. An average of 370 restriction sites per person were analyzed, which tallied to approximately 9 percent of the genome.

Further analyses indicated that the diversity among "Africans" indeed was clearly more than the diversity between individuals within any other group, leading to that conclusion of Africans having been around longer in order to accumulate more mutations. Looking at the ages of certain clusters, the biologists calculated the oldest as being an African cluster that on average was 90,000 to 180,000 years old, followed in order by Asian, Australian, New Guinean, and European clusters. (The European clusters, by this one table, were, on average, but 23,000 to 45,000 years old.)

That, however, was not the whole time story. There was one common ancestral mitochondrial DNA that linked types which had diverged by an average of nearly 0.57 percent. If you interact that with the mean rate of divergence—2 to 4 percent every million years—by dividing both 2 and 4 by 0.57, you find that this decimal divides into 2 percent 3.5 times and into 4 percent 7 times. Take it a step further and divide a million years by both 7 and 3.5, and the result is 142,500 to 285,000 years. That was from an African cluster. (The other ancestral types appeared to have younger ages: 62,000 to 225,000 years.) Scanning the 142,500 to 285,000 range, Cann, Wilson, and Stoneking came up with the date—for that woman who was everyone's common ancestor—of 200,000 years ago. It was the most comfortable inbetween and had a nice round ring to it.

And it would hold up as long as they were right on the 2

to 4 percent rate of mutations, and as long as that rate remained constant.

It was indeed a regular rate, they decided, and so they had just about all the data they needed to go big-time. By 1986, Cann, Wilson, and Stoneking were in the final stages of writing the *Nature* paper, an effort prolonged by Cann's relocation that year to Hawaii. When they got the right computer program going, it had constructed a final tree which placed 133 mitochondrial DNA types at the tips of branches that stemmed from the two main limbs. It was rooted in the middle, assuming equal rates of mutation, which caused a few intellectual battles in the lab because it might leave them open later to criticism. Although there were 147 samples, 14 had essentially the same base sequences as others in the study, leaving the 133 distinct ones. The samples most closely related to one another—with the least difference in their base sequences—were placed together in groupings or "clusters" of tiny branches, and, as Cann explained it, these clusters were connected to others at points closer to the trunk where they shared common ancestors. The branch tips represented all five regional groups: Europeans, Asians, New Guineans, Australians, and some Africans.

The clusters were all on branches that joined at one of those two large limbs—all but seven, that is. These seven were the "Africans." The point at which the two limbs forked apart was the position of the common ancestor whose children had diverged and given rise to the two distinct lines of descent. It lined up with a sequence divergence in the vicinity of 0.57 percent. In other words, it was the position of the faceless, nameless great-grandmother who was described on the tree simply as "ancestor A" but inevitably referred to as "Eve."

There were as yet no fossil bones that could be pinned to Eve. "But we know she existed," said Wilson. "And it is almost certain she was in Africa." There were modern fossils in Africa before anywhere else, he claimed, and so it meant the modern population from Africa had taken over the Neandertal and *erectus* turf in Europe and Asia.

"Africa emerges as our homeland, and the most ancient reservoir of human life," wrote Cann later. "Africa has been a source of continual renewal for our species over the thousands of years of human existence and development."

"When did the migrations from Africa take place?" the three scientists asked in the *Nature* paper. Well, the oldest cluster of mitochondrial DNA types to contain no African members was estimated to be 90,000 to 180,000 years old. "Its founders may have left Africa at about that time," wrote Cann, Stoneking, and Wilson. "However, it is equally possible that the exodus occurred as recently as 23–105 thousand years ago." The latter time range encompassed the youngest date for a European cluster and stretched to the oldest date for a cluster in the population that seemed to have been around longer than any but the African lineage: Asians.

In all the years of paleontologists looking at stone tools and morphologists studying every little nuance of ancient skulls, in all the years of rousing man-ape discoveries and various other indications of protohumans, no one had been able to come up with such a powerful statement about the origins of modern man. There were those who had figured Africa was the origin all along; even the man on the street could have guessed that. But to have a wholly different type of scientific technique arrive so suddenly and settle old questions with equations and precision measurements—that was real drama.

"The Mother of Us All" said part of a front-page headline in the March 24, 1986, San Francisco *Chronicle*.

Sherwood Washburn had his reservations. And Vincent Sarich didn't like it. Sarich believed there was no fossil or archeological indication of such a major movement out of Africa, and he was concerned about the regularity and speed of the mitochondrial "clock." But it was time to move forward, and with Stoneking improving the internal calibration, their confidence grew. That was where the origin would be: in Africa roughly 200,000 years ago.

After years of gestation, after all the internal wrangling and reanalysis, after a year of revisions and collaboration and cogitation, the data finally was sent to *Nature* in neatly packaged form. *Nature* had received the paper on St. Patrick's Day, 1986, but didn't accept it until the following November. It wasn't simply a standard delay. "One gets the impression," says Cann grimly, "they did not want to publish it."

Says Stoneking more stoically: "You always have enemies out there."

The problem was in the peer-review process. A journal like *Nature* sends the submitted paper to two or more experts in the field for an independent evaluation, and when more than one person is called into a case, anything can happen. There were two reviewers in this instance, and while one of the responses was positive, the other was decidedly negative, focusing on the fact that American blacks may not be an authentic source of African mitochondrial DNA.

The delay was excruciating for Rebecca Cann. She wanted to strike first with the African hypothesis and there were competitors out there, also working with DNA, who wanted to do the same, poised, she felt, to steal the thunder. And perhaps they already had. At Oxford, Jim Wainscoat and ten other researchers had published an analysis of nuclear polymorphisms in *Nature* indicating an African origin because Africans possessed a certain gene variation not seen in other populations. Cann felt the Oxford team was discourteous, for Wilson had been to England talking about the new indications from mitochondrial DNA of an African origin, yet Wainscoat "wrote his paper as if he'd never heard that." No mention was made in the Oxford group's bibliography of any material published by Cann, Stoneking, or Wilson. And then there was Doug Wallace, the first to publish clear results that human mitochondrial DNA was maternally inherited and the one who, with Cavalli-Sforza, already had done a large human sampling from around the world. Everyone was breathing down each other's necks. "The delay in publication cost us the momentum we had and the head start we had," groused Cann. "I have blocked out almost all of that. One reviewer was very stodgy. I think sometimes reviewing scientific papers brings out the worst in people. It took a lot of persuasion and a lot of work on Allan's part—because he's the most diplomatic—to get publication of that data."

Wilson was a persistent man, and he had an insider's knowledge of how journal articles were published. He had served on the editorial boards of everything from *Systematic Zoology* to the *Journal of Human Evolution*, and had himself frequently participated as a journal referee. He argued back with force, and *Nature* put a third reviewer onto the case. *That* reviewer generally agreed with the arguments in the article about using a black American sample for their African mitochondrial DNA and, after some rewriting and deletion of a discussion section, the paper was ready for print.

The acceptance, of course, meant another party, and those who were around in the lab had a few daiquiris before bracing for what would turn into a classic and raucous paleoanthropological discussion over the Garden of Eden.

8

It was the world according to Cann. She would be the one called to explain just who this "ancestor A" was. Stoneking didn't possess her knack for communication and Wilson was sliding back into the shadows once the *Nature* paper was out and the coals of controversy stoked. Relying on her training as an anthropologist, along with a unique expertise in genetics, Cann set forth to form a comprehensive interpretation of the peculiar data.

Although, by the cloistered standards of academia, the paper drew quite a public response, it was nothing like a major fossil breakthrough. Not right away at least. Around the time of the *Nature* submission, Wilson gave a lecture for the zoology department at Berkeley and it was there that a reporter from the *Chronicle* had shown up. While biological theory says all species can be traced in principle to one breeding pair, wrote the *Chronicle* reporter, the Berkeley study "is the first to put an approximate date and place on the event for modern humans."

Clark Howell told the reporter that the study "certainly does not violate anything in the fossil record, but it may shake up some people who argue that the transformation to modern humans was more widespread and blurry, and took place over a long period with an extended gene transfer among new and old populations through the Old World." An evolutionist at the University of California at Davis seemed to think people were going to get more than a little shaken up. "First they howl that [Wilson] cannot possibly be right, then a little while later we take a closer look and realize he just might have something."

The article used the expression "African Eve" and led to a small flurry of other publicity. Two days after its appearance, on March 26, 1986, *The New York Times* picked it up and ran a story deep inside its fourth section entitled "Modern Man's Origin Linked to a Single Female Ancestor." The paper made mention of the Oxford group and also of Doug Wallace, who was now at Emory University and whose own mitochondrial DNA results were somewhat at odds with Berkeley.

A second round of media calls—ten or twenty inquiries from newspapers, radio, and television—followed the *Nature* publication on January 1, 1987. A dozen or two calls were not exactly a storm of publicity. Where discovery of the fossil Lucy had made the front page of the *Times,* for example, the newspaper put this story on page D-24, its next-to-last. The *Nature* paper wasn't even drawing as much attention, at this point, as related topics such as Mayan pyramids and ancient astronomy.

But the coals of controversy were slowly growing red, and the world of paleoanthropology, at first baffled into silence, would find its voice when it heard the interpretations that went along with the *Nature* paper, especially the incredible and yet now genetically documented idea that a small band of Africans had stolen the world from anthropology's favorite ancestors. Most irritating was Wilson's attitude that his data was far superior to what paleoanthropologists used, that he was above listening to their objections, and that he didn't owe anyone any more explaining than he felt like. Let them chew on this like they had chewed (and at first had spat out) the 5-million-year ape-human split, which was now so widely accepted. Although it had been two decades ago, Wilson still burned about the initial rejection of his and Sarich's first hypothesis. Increasingly the guru, his résumé more impressive every year, his glistening white hair now falling toward his shoulders, Wilson simply stayed out of reach, assuming the posture of one who was above caring what paleoanthropologists might say. He also affected the air of someone who disdained personal publicity. He'd come forward when there was additional "hard" data, or, more importantly, when there was another hypothesis to propound.

While Wilson's actions seemed like a manifestation of the haughtiness those who knew him often spoke about, they could just as well have been those of a prudent scholar wisely limiting his remarks to avoid the bane of many scientists: misinterpretation. It was also a manifestation of his shy and reclusive personality—a strange mixture of personality indeed. Arrogance and shyness rolled into one. When one broadcast crew stopped by the lab, he scurried into the library, locked himself in, and stayed there all afternoon, according to one of his former lab workers. A crew from the Public Broadcasting System's *Nova* program came around later, and Wilson, according to the same colleague, "went as far into the corner of the office as he could possibly be." Once in a while there were speeches to scientific gatherings—mostly seminars—but that was about it. Any remarks he had to make were contained in the burgeoning pile of technical papers.

Cann was more affable and relaxed. She didn't even care whether people pronounced her name "Can" or "Kahn." She also had an infectious enthusiasm. How could she not be enthused? The vision was enough to stop your breath: of Eve, mitochondrial mom, spawning ageless genes somewhere in deepest Africa, below the timeless, windswept desert. This was no milk-skinned damsel standing daintily next to a rowan tree. That was the stuff of artists like Dürer, who in his engraving *Adam and Eve* (1504) made Eve look like she was waiting to dress for brunch at the country club—a classical beauty if a bit flabby of hip, easily mistaken for the wife of a doctor entering middle age. That was the time-honored Eve. *This* Eve was all grit, tough as tarp to roam the blazing savannah, with no clothes except a hide or two and black (or at least olive-tinted) skin. An Eve from Africa implied an Eve who, contrary to art from the Renaissance, was taut, full of dust, and covered with calluses. The wife of a gazelle hunter.

Yet "ancestor A" would still become known as "Eve": a nickname at once inflammatory, inaccurate, hackneyed, and yet somehow appropriate. Casually bandied about the laboratory or raised at Wilson's weekly progress meetings, the nickname had also been mentioned in a 1983 *Nature* piece reviewing some of the earlier work with mitochondrial DNA. What else but "Eve" to call a single female ancestor?

Still, it was a sobriquet from which both Cann and Wilson would later take pains to distance themselves. Such a name said more than their data was really saying, and it seemed too sensational for a serious, "hard" science like genetics. There was also the matter of simple accuracy. This hypothetical woman was probably not the one and only mother of all subsequent humanity, as the biblical name clearly implied, but simply one woman whose mitochondrial genes got passed along through an endless string of daughters. She was the only woman her age, explained Cann, whose descendants included at least one female in every generation; our oldest known *common* ancestor.

In other words, she was a benchmark with which to gauge times past. And perhaps a benchmark only. Her nuclear genes, which would have been vastly more important in bequeathing to us any physical traits, may have dwindled to the point where, today, they barely exist. Recombination and the contribution of other women would have seen to that.

Which brought up a second major point: the mitochondrial Eve was probably never the only woman on earth. There may have been thousands of others living at the very same time. But *their* mitochondrial lines had gone extinct. They may have sent us even more nuclear genes than "ancestor A," but they hadn't left us a mitochondrial clock. All it would take to extinguish a lineage of mitochondrial descent was a single generation of all sons and no daughters.

Think of it as similar to the way family surnames disappear. Like mitochondrial DNA, a family name is carried forth by only one sex, in this case males, since when women marry they often drop the name they were born with and assume their husband's. If a man has two children, there is a 25 percent chance that they will both be daughters, and if so, his surname will not be carried forth by them. After 20 generations nine out of ten original surnames vanish. After 10,000 generations, only one would be left.

"That's the male analogue of what mitochondrial DNA is doing," says John Avise of the University of Georgia, who corrected some initial misperceptions in the Wilson laboratory. Avise recalls that in 1980, Wes Brown, observing exceptionally low diversity within and among human races, had been led to throw out the possibility—and only the possibility—that all humans may have descended from a single mating pair that lived in a "monomorphic" population (a conclusion reached after consultation with Wilson, who was very enamored of the idea).

"The conclusion they drew was that all humans alive today trace their maternal ancestries back to a single female who was alive on the order of 200,000 years ago. And furthermore, they went on to postulate that this indicated that, at that time, there was probably just a tiny population of humans alive on earth—that the female was the only one around, or that female and her sisters," is Avise's way of recollecting it. "That's when I took issue with it, because in using models for human surname transmission and the extinction of surnames, we pointed out that even if it is true all humans alive today trace to a single female, that does not necessarily imply that she was the only female alive at the time. She could have been a member of a much larger human population and still be the 'lucky' one whose mitochondrial DNA was passed along to us all today. So the only part of their scenario that I took issue with was their conclusion that it was an 'Eve,' that there was one female around in a Garden of Eden."

After Avise's explanation of how one woman could account for everyone's mitochondrial DNA without being the only female from whom mankind actually descended, it became obvious that "Eve," however catchy an expression, wasn't the best nickname for what Berkeley was detecting, but rather one from which the lab should distance itself. This had been realized before the *Nature* paper, yet it was a nickname so alluring as to remain in constant use, finding its way, eventually, into the *Chronicle*'s headline. Wilson pointed his finger at the newspaper for coining the expression and professed discomfort with it. He also seemed to hint that fault for the use of the expression (which he found "regrettable") lay with certain unnamed scientists or journal editors. While he maintained that "there *must* be one lucky mother," he was talking only about the inheritance of a particular type of DNA that constituted something on the order of a mere $\frac{1}{3,000}$th of the body's genes, and genes which, at that, didn't even determine the type of physiological characteristics most people think of when they hear the term "heredity."

"To call her Eve is a bit misleading," Wilson recently told a gathering of scientists. "I feel uneasy about using this term. The term implies that she was the only woman living in her generation and that is very unlikely to be true. We think there were at least thousands of women living at the same time. Our roots trace back to many of the people who lived in her generation. Our nuclear genes were a mosaic made up of contributions from many ancestors in her generation as far as nuclear genes go. As far as mitochondrial genes, we get those only from her."

But the scientist who originated the confusion, according to Cann, the one who egged "Eve" into use, who promoted it, who continued his love affair with the notion, was none other than Allan C. Wilson. He was "trotting around the country calling it Eve and I would get into battles with him about it," Cann says. Brown has a similar recollection. "I remember Wilson tried to get me to put 'Eve' in the title of the 1980 paper, and I wouldn't," he says.

They had known since Brown's days that a population bottleneck, a highly constricted group of survivors that included this particular female progenitor, was not the only scenario, and Cann feared not only the wrong idea being communicated to the scientific world but also the reaction of religious groups. Creationists and Darwinists were fighting tooth-and-nail as it was. There was no use entering *that* fray. Yet on the other hand, says Cann, the "Eve" talk was a welcome diversion. "I was just pleased Allan wasn't talking about two teenagers on a log in Australia anymore."

There were any number of scenarios for Eve. It might well be correct, said Cann, to think of her as a 10,000th grandmother—the 10,000th grandmother, for all we knew, of millions—but not the only grandmother. It wasn't out of the question that she was the better half of a single breeding pair, but this was most likely not the case. She may also have been the very first anatomically modern female, but then again she could have been part of a more archaic group of *Homo sapiens* who served as the most immediate predecessors of modern humans. The population to which she belonged may have been very small, as Berkeley was inclined to think initially, or put another way, a population that, at the point of Eve's existence, had crashed to a very small number for some undefinable reason. In that case she may even have *postdated* the first modern *Homo sapiens sapiens*. Or it might have been, again, that there were many people in her African population—perhaps 10,000 to 50,000, Cann guessed—but, as Avise pointed out, the others' mitochondrial DNA vanished like surnames.

Whatever the case, she was certainly our common ancestor, and that was the most important point. "She's a *mitochondrial* Eve," says Cann. "It's unfortunate as a term if people think of her as a single mating female, who becomes the mother of an entire population, but it is correct in that it draws attention to direct descent that traces back to a single female. She may be the only common ancestor that we all share."

The name "Eve" was mentioned in the very issue of *Nature* that announced the mitochondrial results. In a "News and Views" section where articles summarizing major technical papers in layman's terms appear, the headline on the review synopsizing the Berkeley work was "Human Evolution: Out of the garden of Eden." It was written by the Oxford geneticist Jim Wainscoat, whose own nuclear DNA research also pointed to an African origin but who had made no attempt to put even a rough date on the beginning of anatomically modern man. Wainscoat pointed out that "many Eves have contributed to our nuclear DNA," emphasizing that "we inherit our mitochondrial DNA from just one of our sixteen great-great-grandparents, yet this maternal ancestor has only contributed one-sixteenth of our nuclear DNA." Wainscoat added that a mitochondrial Eve "has contributed little, if anything, to our nuclear DNA," which wasn't difficult to appreciate seeing that the Berkeley group was talking not about a normal great-great-grandmother but of one existing in 200,000 B.C.

Wainscoat went on:

The study of Cann *et al.* represents the strongest molecular evidence so far in favour of the African population being ancestral. The two supporting lines of evidence are the evolutionary tree itself, which clearly suggests an African origin, and the fact that Africans seem to have more mitochondrial DNA diversity than other populations (the population which has been around longest would be expected to have more DNA sequence changes). It seems likely that modern man emerged in Africa and, as discussed in a previous News and Views article, that subsequently a founder population left Africa and spread throughout Europe, Asia and the Americas. Further studies of the mitochondrial DNA types of African populations would be valuable to provide corroboratory evidence for this 'Out of Africa' hypothesis, based by Cann et al. largely on studies of black Americans.

That mention of black Americans was a dig, immediately raising questions about the African sampling. The Oxford geneticist also had this cautionary note: "In considering this striking claim we should bear in mind the words of Thomas Hood in the first stanza of 'A Black Job': 'The history of human-kind to trace/ Since Eve—the first of dupes—our doom unriddled,/ A certain portion of the human race/Has certainly a taste for being diddled.' "

Cann wasn't pleased with Wainscoat's article. ("Very nasty!") You could smell a storm in the breeze. This new hypothesis wasn't going to go down without a few big skirmishes. Already in the *Chronicle* article there had been challenging remarks by the University of Michigan professor, Milford Wolpoff, who immediately questioned the time frame for Eve. Wolpoff simply refused to accept that a population of Africans had spread throughout that continent, wresting control from more primitive types already in existence there, and had done the same in Europe and Asia, edging Neandertals and descendants of Peking Man out of existence without interbreeding with those people whose stock had been around long before 200,000 years ago—in some cases perhaps a million years before the mitochondrial Eve.

Although Wolpoff's objections would come back to haunt them, the Berkeley geneticists contented themselves with taking a few shots back at Wolpoff in lower-profile journals, while seeming to be above the fray in more visible situations. Mainly Wilson ignored the few voices of objection while Cann tried to bring

caution into what quickly became known as the "Eve hypothesis."
"All we can say from the mitochondrial data," she told me, "is that
there was a basic subdivision of some kind or a winnowing down
of lineages that were eventually transferred and that [Eve] may
have been a member of a population that was actually quite large."
There was no evidence, she stressed, when I paid a visit to her new
Honolulu laboratory, that people today retained any of Eve's particular physical features. She may have had nothing to do with our
head and body shape, our hair color, or our eye and skin pigmentation. We had no idea what she looked like. There was even a
chance, theoretically, that her people, though in Africa, were light
of skin.

What Eve looked like for now was a bunch of dark bands
on X-ray or the staggered points and streamers from a penchart
machine. Says Cann: "We can't say anything really about the
morphology that she may have had. The population that she was
representative of may have stayed in sub-Saharan Africa for some
time, and then gone up to some other area and then moved out.
In that movement, at some point, for some unknown reason, the
actual transition from an archaic to an anatomically more modern
man took place. Many nuclear genes from earlier people will survive in modern humans, both male and female, and some of those
could easily come from something that we wouldn't recognize as
Homo sapiens. And most certainly they would have to. The problem is trying to get people to think concretely about what descent
with modification—which is what Darwin was trying to talk
about—really means in terms of the genome. Our genes didn't
come out of nowhere when *Homo sapiens* suddenly appeared. They
were present in some earlier species. And they were reshuffled and
passed on. All I can do with mitochondrial DNA is say something
about the genes that were present and were not broken up in this
mitochondrial suite of characters. But I can't say what was happening in the nuclear genes—I wasn't looking for it. It's unlikely there
would actually *be* a way, given the biology of how nuclear genes
sort, that any of them will survive in the same combination present
in those early species."

The important finding was not of the legendary Eve, but
of a place and point in time where the human race as we know it

today appeared to have been given birth. The "earlier" species Cann referred to was *Homo erectus* or some later form of it, and from the fossil material we knew generally what this primitive human who lived before the Neandertal and who preceded the species we call our own looked like. *Erectus* had been the most successful human up until the modern type, originating in Africa about 1.6 million years ago and spreading around much of the world—including the Middle East and Europe and obviously Asia, where its characteristics had first been defined in Java and Peking Man—until about 250,000 years ago, when it suddenly gave way to a new breed of humanity, replaced totally, it seemed to Cann, by Eve's roaming offspring.

"We don't understand what happened with *Homo erectus*. It's clear that people have this linear view where once you have a population in a certain part of the world it persists, even though we know from demography of modern populations that that's not true. And we know from looking at dispersal patterns in populations in the wild that extraordinarily small numbers of individuals actually do the reproducing for the population."

While some fossil experts postulated that *erectus* had evolved into *Homo sapiens* in more than one place, since there were maybe 1.3 million or so *erectus* on three continents when Eve was born, Cann and her colleagues, reviewing their own analyses and those of others, felt that "if this were the case, the diversity present in the gene pool of present-day *Homo sapiens* should contain deep roots, approximately a million years old. The failure to find any extremely divergent mtDNA lineage in surveys of more than 500 Asians makes it unlikely that Asian *Homo erectus* contributed much to the gene pool of anatomically modern *Homo sapiens*."

It meant too that the Old World had been settled at least twice—and by two different human species. First there were the *erectus* who also originated in Africa and went as far as Java and Beijing, evolving eventually into archaic *Homo sapiens* such as the Neandertals when they finally reached Europe. Then, 100,000 to 200,000 years ago, the first modern *Homo sapiens sapiens* rose from a uniquely advanced band of archaic *sapiens* in Africa and set out for a journey very much like that of earlier, more primitive humans, heading for even more far-flung parts of the world, and replacing the unevolved descendants of *erectus* without breeding with them.

Apparently, Eve's offspring weren't very attracted to the old fogies living in caves at places such as Zhoukoudian. Their overpowering of *erectus*, their usurping of *erectus*'s niche, left no intermingling genes. The replacement was total, presenting little or no admixture, but this "conquering" of the world by Eve's descendants, says Cann, was not the common concept of invading populations. The takeover may have taken 3,000 generations to complete. It was more like *outliving*—possessing some adaptive advantage that put them over the top—than invading.

"I think that human populations were constantly branching off and trying new things—biological experiments," Cann explains. "Most of them didn't make it. It may have been just random chance that this group from Africa did essentially make it and make it for a long period of time. It might have been benign environment, it might have been superior cultural advantage— who knows? Something."

They'd had to have come up with an advantage that gave Africans—Eve's band—a crucial edge. Something had caused them to be significantly different from old *erectus*, whose lives, felt Cann, had been "nasty, brutal and short." The salient point was that Neandertals and Peking Man had nothing directly to do with our heritage, and there could be no more than a few paleoanthropologists in the world, no matter how Clark Howell soft-pedaled it, who would remain sitting and take no notice of *that* claim. True, there had always been a good number of theorists who believed a small population had taken over the world, but this assertion from Berkeley was much more extreme and was based on new and unprecedented evidence. *Complete replacement. No admixture. Erectus in Asia merely going extinct.* A lot of views would have to be erased in one fell swoop. The followers of Weidenreich would be up in arms. Even Huxley would have raised a quizzical eyebrow.

Yet Cann was never overly concerned with what the old school thought. Genes were genes. It was her contention that the first modern humans had left Africa somewhere between that 90,000 to 180,000 year period mentioned in the *Nature* paper as the age of the oldest cluster of mitochondrial DNA types containing no African members. In Cann's estimation, the migrating Africans—soon to evolve, of course, into non-Africans—most likely went to northern Africa and on to the Middle East. Perhaps they left because kids were getting malaria, says Cann, or because the water holes had grown foul, or simply because a rising population was creating tension.

The movement wasn't a massive march but more likely a gradual fanning out. Every generation, a clan may have expanded outward until suddenly—there was the Mediterranean. Although it was more than 2,000 miles from Tanzania to the Middle East, that isn't so far when one considers expansions and migrations that occurred over the course of 100,000 years. Movements of just 2 miles per generation could tally to more than 8,000 miles over such a period of time.

If these *Homo sapiens* ran across populations of *Homo erectus* in the Middle East or later Asia, there had been, of course, little or no sexual interaction. Perhaps the unevolved *erectus* were too hairy to appeal sexually to a population that had discarded the old *erectus* traits and now was taking over the territory, or perhaps they simply turned off Eve's great-granddaughters with their funny, sloped foreheads and uncouth browridges.

Or maybe it was just xenophobia. As Cann points out, aboriginal populations are known for their cool reaction to outsiders. Their world is defined as "us." They may consider outsiders unreal or some sort of animal. Fear and scorn are the reactions that greet someone who looks different from those in a specific clan.

It was Cann's feeling that instead of waging warfare and killing off all the *erectus* and Neandertals, the out-of-Africa *sapiens* had simply out-reproduced and replaced older human forms, bringing more children to maturity through their more modern, sophisticated, and apparently efficient lifestyles. Language, she believes, allowed the *sapiens* a competitive edge, especially among youngsters, "because they're better at remembering they're not supposed to play with those big wooly animals and are supposed to remember how to get back when they're out wandering in the woods or better at making connections of why not to do something when told not to. Communication by spoken word was probably an explosive stage."

With language, feels Cann, the *sapiens* were able to communicate the locations of watering holes (when drought struck one place) or better warn one another of lurking dangers.

So an adept and isolated band evolving out of *erectus*-like ancestors south of the Sahara had taken a critical leap that was to define modern man. Cann points to fossils of anatomically modern humans found in South Africa, East Africa, and the Middle East at time periods that coincide fairly well with the mitochondrial data, which, she says, indicates that although more primitive men had found their way into other parts of the world long before Eve's time, the first *Homo sapiens sapiens* had arisen only in Africa. The land of apes, as Darwin himself had predicted, was also the land of man.

9

"IN EACH GREAT REGION of the world the living mammals are closely related to the extinct species of the same region," wrote Charles Darwin in *The Descent of Man* (1871). "It is therefore probable that Africa was formerly inhabited by extinct apes closely allied to the gorilla and chimpanzee; and as these two species are now man's nearest allies, it is somewhat more probable that our early progenitors lived on the African continent than elsewhere."

It was Sarich and Wilson's old rival, Elwyn Simons, formerly of Yale, who discovered the most primitive of the extinct apes, and yes, the place was Africa. This was a fox-sized creature in the Fayum of Egypt known, after the country in which it was found, as *Aegyptopithecus,* discovered in 1966. The obsolete primate, often described as half-monkey and half-ape—monkey legs and ape teeth, possessing a tail—was really, in Simons's view, mostly ape, with more complex teeth than a monkey's, heavy, deep cheekbones, and frontal crests on the forehead that converged to a sagittal crest like what is seen in a gorilla or male orangutan. It probably ran along tree branches instead of swinging below like a classic ape, and it is estimated that this specimen, found below a layer of basalt sixty miles from Cairo, is 32 million years old, from a period known as the Oligocene.

"Most of the similarities in the whole of the skull make it look like an ape," says Simons. "It doesn't look like a monkey. Monkeys have more delicate cheekbones today. People think the

119

oldest transitional form should look like monkeys because monkeys are hypothetically at an earlier evolutionary stage, but as I tell
my students, what you find from the past is never what you expected."

No one can be sure what the picture looked like so long
ago. Remarkably, science has not yet been able to settle the fundamental question of which came first, monkeys or apes. There are
those who believe *Aegyptopithecus* came *before* the split between
lines leading to monkeys and those evolving into apes. Others
believe the eight-pound primate existed just after that split. Only
one thing is certain: in most instances judgments and deductions,
never short of imagination, are made on meager evidence—as if
by touching a fossil its past can be resurrected and materialized like
the shimmering image in a crystal ball. "In view of the very fragmentary nature of the fossil evidence," writes Martin Pickford, a
paleontologist, "the reconstructions probably reveal more about
the scientists who made them than they do about the species they
purport to represent." Indeed, paleontologists have been known
to describe everything from diet and hunting patterns to intelligence, sexual conduct, and walking ability based upon bits of bone
that could fit into a shirt pocket.

"The thing about *Aegyptopithecus* is that it's so primitive
that it's really unrealistic to refer to it as having many resemblances
to living things, whether monkeys *or* apes," says David Pilbeam,
disagreeing with Simons, his former partner. "When he says it's
apelike, I think that's an ambiguous use of the term 'apelike.' If
you were looking at it running around in a tree, you'd think of it
as being a cross between a lemur and some kind of New World
monkey. It's just a very generalized, primitive animal."

Primitive indeed. These are the life forms that supposedly
sent organisms onto the path of becoming human (or at least
becoming modern apes), but caution has to be urged at every step
of the way when considering a history of such daunting distance
and elusiveness. Before the monkeys or apes were those rodent or
squirrel-like prosimians (later looking more like little koala bears)
which already had spread around the world, including into the
Americas, and so the migration patterns and points of origin for
any subsequent creature are open to question. Teachers of evolu-

tion have presented the public with a picture of ancestry that is much simpler, more linear, and far more certain than reality dictates.

For one thing, there is always a problem with dates, which are constantly subject to change. For a long time *Aegyptopithecus*, that first "ape," was thought to be 4 million years younger than what is now accepted as its true age, based on redating of the Egyptian samples. Back in the Oligocene were creatures whose entire physiques and habits, as Pilbeam points out, were unlike anything alive today, and no one is truly sure just which of those primitive creatures—if any—truly fit into our ancestral scheme. Although he doesn't himself subscribe to the theory of *Aegyptopithecus* being ancestral to all subsequent higher primates, Simons, who is to apes what Sarich was to molecules, has never determined in his own mind just when monkeys and apes went their separate ways.

But Simons as well as many other zoologists and paleoanthropologists seem to have reached a consensus that, after *Aegyptopithecus*, the next major ape on the hypothetical progression to humans was probably a creature called *Proconsul*, which once more pointed to Africa as the primordial melting pot. Named whimsically after a London chimp, "Consul," that entertained vaudeville audiences by riding a bicycle and smoking a pipe, it was discovered in 1927 by H. L. Gordon, a settler in western Kenya digging around in a limestone quarry. What Gordon found was an upper left jawbone known as the left "maxilla." Though that may not seem much to go on, scientists studying the fossil determined that it belonged to a new genus ancestral to the chimpanzee. In 1948, Mary Leakey, who rarely was to be outdone, found her own specimen of *Proconsul* on Rusinga Island in Lake Victoria: not a simple maxilla but instead an actual skull.

A skull is much more to go on than a maxilla, of course, and staring at this remarkable, squashed little skull is the morphologist's equivalent of a mystical experience. In the years since *Proconsul*'s discovery, anatomists and the anthropologists with whom they associate have formed the picture of an ape that was about the size of a baboon, with a cranial capacity of 167 cubic centimeters, which is half that of a chimpanzee's. Other fossil discoveries have shown that *Proconsul* had ankle bones that were slender and comparable to a monkey's but a big toe that was robust and like an ape's, lumbar vertebrae that resemble a gibbon's, and shoulder joints like those of a chimp. By this time there was no more tail. *Proconsul* is thought to be 18 million years old, from an ape-rich period known as the Miocene. It may have been the rootstock for any number of later anthropoids.

"Without saying whether it was directly ancestral or not, *Proconsul* represents one of the very earliest apes and comes at a time in the Early Miocene just after the split in the lines leading to the living monkeys and living apes," says Pilbeam, who describes *Proconsul* as a fruit-eating, tree-dwelling animal which was distinctly different in size depending on whether it was male or female. It's a characteristic shared by humans. During its existence the earth was racked by a series of momentous geological changes that may have abetted migration of such primitive apes from Africa into Eurasia. For one thing, continental drift linked up the Middle East and Africa to Asia during this age. Europe and Asia were mainly woodland with milder winters than are experienced today, and such a climate proved hospitable to a close relative of *Proconsul*'s known as *Dryopithecus*, which dwelled, among other places, in southern France. Time and again, through the epochs, Africa seemed to have spawned new genera only to project many of them outward to form still other genera or species in the reachable and temperate parts of the world.

One of the most exciting realizations he has had, says Richard Leakey, is that the Miocene has a far greater complexity of primate forms than anyone suspected. "The eastern side of Kenya has an enormous potential for Miocene research, untouched," Leakey says. "We may well be able to trace back a savannah ape much earlier than we hitherto thought. We may find the underpinning of the hominid story is much more interesting and complicated than the traditional Miocene evidence from western Kenya has suggested. In the last three years there have been found four new genera of Miocene hominoids, doubling the genera."

Dryopithecus, which resembles *Proconsul* but is associated with Eurasia, was followed by two more primitive apes that are so closely aligned they may be of the very same genus. These are *Sivapithecus*, the remains of which have been found in both Asia and Africa, and *Ramapithecus*, the 7- to 15-million-year-old species once thought by scientists, especially Simons and Pilbeam, to be the link between apes and humans, based mainly on the discovery of two jaw fragments. That interpretation was the one eradicated by the molecular findings of Sarich and Wilson, who seemed to have shown that *Ramapithecus* lived *before* the ape-human split, along with discoveries by Pilbeam in Pakistan indicating, as we saw in Chapter 4, that *Ramapithecus* was not a direct ancestor of man but rather in the ancestral closet of the orangutan. Though there are those who still maintain that the orangutan, which is confined to small parts of Asia, is perhaps man's closest animal relative, most paleontologists believe that the line leading to the orangutan went a separate way about 10 million years ago while other ancient apes, on another side of the tree, evolved into those largely undefined creatures that served, most likely in Africa, as precursors for the more closely related gorillas, chimps, and protohumans.

After spreading from Greece to China, the apes such as *Sivapithecus* and *Ramapithecus* went extinct about 7 or 8 million years ago. One of those which remained around—on the branch that would give us the orangutan—was a giant ape in Java, China, and Vietnam that lived mainly during the period known as the Pliocene. Called, logically enough, *Gigantopithecus*, this creature had jaws and teeth larger than the largest modern gorilla and, by those who refuse to believe it is fully extinct, has been pointed to as a candidate for the legendary Yeti (better known here as an "abominable snowman") of the Himalayas.

So what we have is a tangled and unresolved notion of the ape picture, and if there is an accurate evolutionary tree for the period leading to the all-important evolution of gorillas, chimps, and humans, it would look not like a large oak but rather a series of intertwining vines. *Aegyptopithecus* and *Proconsul* and another ape called *Kenyapithecus* probably wound their way into our lineage, with *Sivapithecus* and *Ramapithecus* branching off on their own, *Gigantopithecus* forming another off-branch a while later (on the same side as other Asiatic apes such as the orangutan), and the last truly primitive apes remaining a mystery, for there are few good fossils from the Late Miocene and none that definitively prove any particular interpretation.

In this head-spinning scenario of the past, with argument over which genus or species many creatures truly belong to, with genera switched from one lineage to another like calls on a phone board, replete with dozens of species names, with controversy *still* lingering about whether *Ramapithecus* should be reinstated as a direct ancestor and whether *Aegyptopithecus* was an ancestor after all, with age estimates that can vary by 10 million years and overlap each other, skewering any attempts at a simple family tree, with questions even about whether Asia might have been the true origin for ape ancestors (though Africa remains the most probable), there is only one thing for certain: Out of this formidable chaos, out of the melange of creatures that looked something like an ape but also something like a monkey or a fox or perhaps even, at the extreme end of size, the abominable snowman, came the chimp, the gorilla, and "hominids," or prehumans. Those were all from Africa.

As discussed in previous chapters, it was the imposing and bearded Vincent Sarich, along with the much shorter and more aloof but equally mischievous Allan Wilson, who first insisted, based on protein distinctions in the 1960s, that the lines evolving into gorillas, chimps, and man split from each other relatively recently—about 5 million years ago, give or take a few million years. Certain authorities such as Richard Leakey believe the dates Sarich and Wilson gave for the split are too young and simply wrong; nonetheless, their findings caused a wholesale revision of ape phylogenies and converted even some of the most ardent fossil hunters, excluding the illustrious Leakey, to their way of thinking. One of them was David Pilbeam, now at Harvard, who at first had looked at the molecular data and wondered "how in the name of Heaven anyone could believe it," but who, in 1984, was to write that "the molecular record can tell more about hominoid branching pattern than the fossil record does."

Pilbeam's conversion to molecules came after his reputation had been seriously challenged, along with Simons's, by the warfare with Wilson and Sarich over *Ramapithecus*, which the two fossil men from Yale insisted was a "hominid," or man-ape, but which Sarich insisted, on the basis of molecular dates, was nothing more than another ape. Sarich, for the time being, had won that war. He had done it in what Simons described as that "swaggering" and uncompromising fashion, and after more than a decade there were still bitter feelings (see Notes). Now both Simons and Pilbeam would have peripheral and unexpected roles to play in yet another battle over ape phylogeny that embroiled the Berkeley lab just as it was encountering its first reactions over the Eve findings.

At issue was the conundrum of just how apes led up to that first hypothetical human. One burning question that needed answering before anyone could claim even a crudely comprehensive knowledge of ape evolution (and thus the rise of man) was the roles and relationships played by the most modern of apes: chimps and gorillas. Although Morris Goodman had long ago noticed the molecular closeness of all three species, and although Sarich and Wilson, in the mid-1960s, had seen surprisingly striking similarities between humans and chimps (which were about as close, said Wilson in 1975, as sibling species of fruitflies), no one had resolved what became known as the gorilla-chimp-human "trichotomy." That's to say, no one knew just when each went their own way, and thus no one knew which two of the three species were most closely related.

It was a crucial question, and the general supposition, unsurprisingly, was that the chimp and gorilla are closest. They are both apes. Both use their knuckles to walk. Both are associated with the jungle and are covered with thick, coarse hair. Neither one, in its natural habitat, is able to form words with the vocal cords. To the general public, which often fails to differentiate between primates, both looked, in short, like overgrown monkeys and thus bosom relatives.

Pilbeam, convinced by now that molecular data could very well solve the puzzle, helped to persuade two colleagues at Yale, Charles G. Sibley, one of the world's most prominent ornithologists, and Jon E. Ahlquist, a collaborator in Sibley's lab, to investigate the relationships between chimp, man, and gorilla. He went to these scientists for a very good reason: Sibley and Ahlquist had been reconstructing the family tree of birds using a new and powerful technique for analyzing DNA. Pilbeam wanted them to apply that method to compare apes with humans. In the end there would be spectacular results and what can only be described as spectacular intellectual warfare between Sibley and Ahlquist on one side and the previous superstars of primate analysis, Wilson and Sarich, who obviously were taken aback by the Yale team.

The technique is known as DNA hybridization, by which the DNA distances between various species are measured with the use of heat. First, heat is used to separate the double-stranded DNA into single-stranded fragments, and then fragments of one species are combined with the single strands of another to see how they bind with each other. The mismatched strands—for example, chimp DNA with that of a human—combine only to the extent that they "recognize" each other's base sequences. The more similar are two species, the tighter they bind to each other. This similarity can be measured, in statistical fashion, by again applying heat but this time to separate the *hybridized* strands and record what temperatures were required for the separations. The more heat that is needed, the more closely related are two species.

In this fashion Sibley and Ahlquist, during the course of one ten-year period, had conducted a phenomenal 25,000 DNA hybridization tests on 1,600 species of birds, revising and refining concepts about their relationships. For instance, they had discovered, contrary to popular conception, that mockingbirds are most closely related to starlings, which had previously been aligned with crows. They dated the divergence between various lineages on the basis of the logical assumption we have seen in other work: that genetic relationships among species reflect their evolutionary past, and that genetic differences were related to how long it had been since two lineages shared a common ancestor.

In 1984, Sibley and Ahlquist reported in the *Journal of Molecular Evolution* that, as Pilbeam had urged, they had applied their technique to gorillas, chimps, and humans. The DNA was from the nucleus, not the mitochondrion, and the technique allowed a larger if cruder purview than DNA sequencing. Sibley says he expected, as most everyone else did, that chimps and gorillas would have had more in common with each other than either of the two apes would have with humans. When the results came in, he had a simple reaction: "I was astonished."

So was the scientific world, for Sibley and Ahlquist declared, in their report, that human DNA combined slightly more securely with chimpanzee DNA than either did with the genetic stuff of gorillas, indicating that humans and chimps were closer to each other than to the gorilla. They repeated the conclusion in 1987, adding further data. Man and chimp, again, weren't cousins, said Sibley and Ahlquist, they were evolutionary brothers.

This ran not only against the popular grain but also counter to previous molecular conclusions. Although Wilson and Wes Brown had not cracked the man-ape riddle, they had found somewhat more of an affinity between gorillas and chimps than between chimps and humans. That work had been done, of course, not with hybridization techniques but with the alternative and perhaps even competing methodology based upon mitochondrial DNA.

More, while they all agreed how phenomenally close all three species were, Sibley and Ahlquist devised a time scale that was a bit different from what Sarich and Wilson had produced with the proteins. Apparently this didn't sit well at Berkeley, where reputations had been made on the time frame of 5 million years. The Yale biologists saw man and chimpanzee splitting from each other 6.3 to 7.7 million years ago, while gorillas had separated from their line about 2 million years earlier. In other words, at the extremes of their two different ranges, Sibley and Ahlquist's estimate of when gorillas left the human line was twice as long ago as Wilson (who told me he now uses 5 million plus or minus 2 million years, which could mean 3 million) and Sarich had claimed.

Did this have any bearing on the Eve research, which had used the chimp-human divergence of 5 million years as a benchmark to devise the all-important mutation rate? The distance between humans and chimps, when the DNA was heated apart, gave a value (expressed as $\Delta T_{50}H$) of 1.8° for humans and chimps, 2.2° for chimp and gorilla, and 2.3° for human and gorilla, granting even those unfamiliar with the mystifying world of heteroduplexes an idea that the smallest differences were between chimpanzees and humans.

This was news; big news. The three-way split, one of the most immediate and relevant of all questions about ape ancestry, had finally been solved—a feat that had eluded Berkeley. Further-more, Sibley and Ahlquist had figured out when other modern apes had risen from the tangle of those primitive creatures. They reported that gibbons branched off from the line leading to hu-mans about 20 million years ago (twice as long ago as Berkeley had calculated), orangutans diverged 13 to 16 million years ago (not 8, as Sarich and Wilson originally claimed), the gorilla line split off somewhere in the vicinity of 9 million years ago, leaving chimps and humans together until they branched from each other 2 mil-lion years later.

Sibley and Ahlquist's DNA technique was impressive. It certainly seemed to have higher resolving power than Sarich and Wilson's proteins, which were coming under increasingly skeptical scrutiny. Morris Goodman, whose very work served as the precur-sor to Sarich and Wilson's research, had, by 1982, expressed his view that individual proteins were very poor evolutionary clocks because they evolve at markedly non-uniform rates! (This was naturally a big blow to Wilson and Sarich, whose careers had been built on the regularity of molecular clocks, and who thought that war had been won.) Moreover, if the protein clock was not so regular, then the same might well be true of the mitochondrial one. To make matters worse, in the very journal where they announced their new dates for the ape-human split, Sibley and Ahlquist them-selves expressed the feeling that clocks based upon proteins were bound to be "more erratic and 'sloppy'" than the DNA clock, which is averaged across the entire genome."

For his work, Sibley, who retired from Yale to set up a lab out west, was elected, in 1986, to the National Academy of Sciences.

Not only were Sibley and Ahlquist Johnnies-come-lately grabbing the glory and stepping on Berkeley turf, but they were being indirectly and perhaps not so indirectly critical of the method of albumin comparisons Sarich and Wilson had ridden to fame in the hotly competitive world of evolutionary research. In Ahlquist's estimation, his and Sibley's work had "eclipsed" what Sarich had earlier done with albumin. Since Sibley and Ahlquist were at Yale, it all harked back to the Yale-Berkeley competition between Simons and Sarich over *Ramapithecus*. Indeed, who else was listed as one of the sources for Sibley and Ahlquist's tissue samples but the man whose reputation had suffered most at the hands of the earlier protein work, and whose *Ramapithecus* had been dethroned as a hominid partly because of Sarich's time scale for the ape-human split—Elwyn Simons.

There was another contest at hand, and this time Yale had taken the lead. The new war came just as the Wilson lab was becoming preoccupied with the Eve research, but there was the past to protect, and apes, as Eve's rootstock, remained extremely important. There was also the "clock." Sarich acknowledged that no other single development in molecular systematics "so captured our attention as their claimed resolution of the longstanding trichotomy" linking humans to chimpanzees. He was also to acknowledge, no doubt biting his lower lip, that Sibley and Ahlquist had developed "what is, without doubt, both the best and most cost-effective approach to molecular phylogenies," describing the hybridization technique further as a "magnificent achievement."

But here was the catch: Sarich also claimed Sibley and Ahlquist had used a statistical method that was "fatally flawed," had failed to reveal all their raw data, and far worse, that the two biologists had "fudged" their numbers to get the dramatic results they did. There was a thin line between claiming that and accusing the Yale biologists of fraud. It nearly immediately shaped up to be about as nasty a controversy as there had ever been in the long and tortuous history of research into ape evolution.

It was an unusual fracas not only because of the tone of debate, but also because Sibley, by his own reckoning, was a longtime acquaintance of both Sarich and Wilson's, albeit not always in agreement with them. Sibley says he was the one who had helped Wilson get his first NSF grant, and in 1969 he had taken both Sarich and Wilson on an expedition to New Guinea, a place that was later to figure prominently into the Eve research. It was during that expedition, right after Sarich and Wilson had first divulged their albumin results, that Sarich and Sibley, who was not so immodest himself, first found themselves at loggerheads. Two egos that size were too much for the boat. At the time Sibley simply didn't believe in any such thing as a molecular clock, and it so rankled Sarich that he never forgot it.

Now, two decades later, in the midst of the revelations about Eve, Sarich was on the attack. He was going to remain a loud voice in the question of ape-human divergence. He was going to bring Charles Sibley, old and aging anyway, to his knees.

"We know that some of the hominoid-primate data that were published in the latest journal article were fudged—changed to look better," Sarich told me. "It's not simply a matter of interpretation. There has been, apparently, even on the primate stuff, wide-scale enhancement of the data. 'Fudging.' That's the term I like. They've been claiming more than the data says."

Sarich was saying that the published results of Sibley and Ahlquist did not properly represent the raw data, and that therefore their claim to have resolved the relationship between humans, gorilla, and chimp was not supported by their data; they hadn't outdone Wilson and Sarich after all. Sarich was joined in the assault by Jon Marks of Yale (showing that it was really not so much a clash between schools as it was a personal confrontation). Like Sarich, it was difficult for the outsider to tell if Marks had an ax to grind or if it was simply a matter of valiantly exposing what they saw as a breach of sacred scientific law. A short while before Sibley and Ahlquist published their startling find, Marks had presented a thesis on chromosome banding that was directly contradicted by the Sibley and Ahlquist results, for Marks's work supported the more orthodox notion that chimps and gorillas were closer to each other than either was to humans.

When I asked Marks whether Sibley and Ahlquist were guilty of actual falsification, he replied, "That's what I'd call it. In fact it's what I *do* call it. There are people who would reserve the word 'fraud' for people who don't even do the experiment but write it up as if they'd done it. These guys had done the experiments as far as we know. They simply misrepresent the results. If we want to call that fraud, quite frankly, I would at this point."

Marks went on: "That is to say, they tampered with their data in such a way as to pull conclusions out of their data that aren't *inherent* in their data. We asked Sibley and Ahlquist if we could see their documentation and were simply told no. Hybridization gives crude data, and we're not sure it can distinguish close relatives and close branches. They published a second paper in

1987 in which they actually gave their experiment numbers. We compared the numbers we had of the same experiment with the values they published and found 40 percent of them didn't match. So there was clearly something going on that they weren't telling us about."

Echoes of the Piltdown affair, back in the 1920s, when a "prehistoric" man discovered in England was found—forty years later, after phylogenies had been redesigned to include this important new discovery—to have been a hoax in which someone had cleverly combined a modern human skull with the jaw of an orangutan, giving it a dramatic and primitive look. But this was much different. It wasn't a hoax, but more a matter, agreed many onlookers, of Sibley and Ahlquist perhaps taking too many liberties—and making too many judgment calls—with the data. One of their co-authors had withdrawn from their 1987 paper because the numbers Sibley and Ahlquist were coming up with did not match his. The work was at the least sloppy and, when exposed by Sarich and Marks, highly embarrassing. Ahlquist explained that the "tampering" with data was simply to correct certain values, discriminating bad temperatures of melting curves from good ones. In essence the adjustments were made to get rid of artifacts that had resulted from a technique of radioactive labeling.

"We were going to explain it in a later paper but it was around the time Sibley and I were both moving to different schools and we forgot and let it slip by," says Ahlquist. The reason the raw data was not readily available for outside scrutiny, he says, was because it was not computerized. As Sibley pointed out, a follow-up study by an independent researcher, Jeffrey Powell of Yale, who was urged to repeat the experiments by Marks himself, got results that had to be deeply disappointing to both Marks and Sarich, for they vigorously and rather unambiguously confirmed what Sibley and Ahlquist had found. In the meantime, Goodman at Wayne State was finding that his own DNA analyses based on nucleotide sequencing of globin genes also showed that humans and chimps shared the fewest differences, as did the work of other researchers with no ties to Sibley or Yale.

Although critical of Sibley and Ahlquist for not reporting corrections made in the data, hardly anyone else thought what they had done amounted to fraud. Emile Zuckerkandl, the man who, with Linus Pauling, had proposed the very idea of a molecular clock and served as the editor of the journal which published Sibley and Ahlquist's results, saw "no such thing" as blatant fraud in their work, and Roy Britten of the California Institute of Technology (who was instrumental in devising the hybridization method) told me the reason for the Marks and Sarich attack was not based so much on scientific grounds as "strategic" and "political" ones.

In coming to Sibley and Ahlquist's defense, Britten speculated that perhaps Sarich was mad because Sibley had been elected to the academy, while Sarich never had been bestowed the same honor for similar work. Britten was the first to see the raw data and in fact was the one who gave some of it to Marks and Sarich. Explaining what Ahlquist had done, Britten says, "He was using judgment based on years of measurements, and he went a little overboard on that. [In] the particular case on which the attack was made, what happened was that a control failed. Ahlquist had the choice of throwing out the point or correcting it to the expected TM difference. That changed 40 percent of the data that got into Marks's hands."

Britten further claims that Sarich and Marks themselves used "numerical trickery" in *their* written attacks on Sibley and Ahlquist. Britten says the allegedly deceptive or defective data was contained in widely circulated "attack" papers Sarich and Marks had devised after scrutinizing only a small portion of the raw hybridization numbers, and Britten helped stop one of those scathing (and in his opinion badly flawed) manuscripts from getting into the *Journal of Molecular Evolution* despite the efforts of Allan Wilson to have the attack printed. Wilson, who served as an associate editor of the *Journal,* was in charge of the review process for the critical Sarich paper despite his obvious closeness to Sarich going back two decades. In other words, Wilson, until Britten intervened, was organizing the judgment of a paper produced by someone who maintained a desk in his own lab.

On the other hand, Britten, who pioneered the technique, was only included as a commentator on the Sarich submission when he learned of the critical paper and requested a copy for review, speaking directly with the editor-in-chief, Zuckerkandl. "What Wilson did was try to get that [Sarich] article in print in a hurry," says Britten. "This paper had been in Allan's hands about two weeks, as I remember it, which was very quick for four reviewers to get their reviews in, when came a message from Allan asking would I please Fax my review back to him because they were ready to move. That I've never seen in that short length of time." Adds Britten: "There's bad blood between Wilson and Sibley going way back." Wilson, in Britten's opinion, was getting downright "mean" and showing "very bad behavior."

When told of Britten's accusation of "trickery" in *his* paper, Sarich responded that "Britten's comments are disingenuous. We were the first to present the raw data. If we were playing number games, everyone would be able to see it." Nor did Sarich believe Wilson had rushed the article into print. "We sent it in during April, and it wasn't accepted until September," says Sarich.

The episode was a revealing one in demonstrating how supposedly objective scientists often interpret the very same material in a way that fits their preconceptions, ambitions, or earlier pronouncements. It also demonstrated the force with which the Berkeley lab operated, and its shrewdness at conveying its standpoint in strategic forums. Just as Simons and Pilbeam had suffered for disagreeing with Sarich and Wilson about *Ramapithecus,* now Sibley and Ahlquist, who were in effect competitors, were also feeling intense heat—rightly or wrongly.

The Wilson lab took no prisoners. The consequences, both personal and professional, had the potential of running deep. During dinnertime conversation in his own home, Sarich raised the possibility that Ahlquist might commit suicide because of the exposé. Although Sarich's paper attacking Sibley and Ahlquist had not yet gone through the review process (and was rife with unusually harsh language), Wilson, who also served as a reviewer for NSF grants, had attached a copy of the damaging manuscript to an assessment of a research grant proposal submitted to the NSF by Sibley, who was now in California setting up a lab at nearby San Francisco State. Sibley was denied the grant. And so Wilson may have helped kill Sibley's request for research funds— from the very same organization that Sibley said he had helped a young Wilson obtain some of his first money from.

For his part, Wilson viewed the counterattack against Sarich and Marks as a "general smokescreen of *ad hominem* attacks" that avoided the real scientific issues. "And the smokescreen is still going on," says Wilson, who normally shunned journalists but felt compelled to comment on this particular issue. "In regard to what I did, I saw the same reaction to me. The best I could do is draw their attention to an unpublished manuscript. I presented evidence on which judgments could be based." He felt that attaching the unpublished, unreviewed, and strident Sarich paper with his assessment of a Sibley grant proposal was not only proper but the only right thing to do. "It would seem to me to be an act of censorship not to include it," he says. "I was doing nothing different than what other people do, except I was being honest."

Bad blood indeed. Marks thought Wilson's handling of the grant proposal justified (and violated no specific rules), while certain other observers, according to *Science*, considered Wilson "to have acted improperly." When I saw Charles Sibley during a conference near Monterey, California, in June 1988, he looked very much like a broken man. Dry-mouthed and visibly upset, the aging Yale ornithologist, suffering from a heart ailment that necessitated surgery, barged into a seminar in which Sarich was leveling another attack. He called Sarich's critique "slanderous" and later, in a conversation with me, described Marks as "execrable in his behavior." It had gone, he said, "beyond civilized discourse" and was all caused simply by "experimental error."

Even Sibley's bird work was being called into question, and he lamented the fact that he could not raise grant money for the laboratory he was trying to establish, forced to spend $26,000 a year of his own money to keep his lab workers going—half his salary, he claimed. "Very nasty things" were done not only by Sarich, he said, but also Wilson. "It has been very painful. Our numbers were values that were calculated and then corrected for obvious aberrations and those are the kind of corrections that are applied routinely." The mistake was in not openly indicating the changes. The mistake may also have been in meddling with hominoid evolution, which, as we will see again and again, causes the immediate stirring of the most primal emotions and rancor. "I would be happy to just erase the hominoid stuff from the record, because I don't really care about it," says the bird expert.

But the findings, in conjunction with Wilson and Sarich's earlier work, have already changed the face of evolutionary conceptions, and independent results indicating the same conclusion apparently remain intact. At the very least Sibley and Ahlquist provided another and more widely quoted point in time during which, rising from the half-apes and half-monkeys, chimps and gorillas split from each other and also from the line leading, in Africa, where the clement weather and dynamic environment nurtured such a cacophony of creatures, to the first hominids, or "man-apes."

10

FORTUNATELY FOR HER, Rebecca Cann was halfway across the Pacific from the fireworks at Berkeley, busy running her own lab on the island of Oahu where certain native birds were going extinct (much to her dismay) and where cresting waves fairly resonated with the echoes of evolution.

Although there already had been critical letters to *Nature* on the Eve hypothesis (and forces of more criticism were gathering like those morning mountain clouds), Cann's thoughts remained on the essential question of precisely how an organism as wondrous and different from animals as *Homo sapiens* had risen from the ape past to conquer the third planet from the sun.

What she ruminated about were those first exquisitely primitive beings that had separated from the chimp and had gone their own way. "These," she figured, "were kind of wild animals. And we were the domesticated version of them."

In the words of one paleoanthropologist she knew, Alan Mann, formerly a student of Sarich's and now a professor at the University of Pennsylvania, what walked the earth before the first human were creatures "different from anything that has been seen on the planet or is on the planet today." It was hard to say much more than that. There was a gap in the fossil record of about 4 million years that obscured the period before, during, and just after the supposed split between gorillas, chimps, and prehumans. All there was to go on were questionable bits and pieces in Ethio-

137

pia, where several fragments of skull and the upper portion of a thigh bone nearly 4 million years old were found in the Middle Awash Valley, and more scraps of fossil in Kenya: an upper arm bone from Kanapoi at the south end of Lake Turkana that might also be 4 million years old, part of a jaw from Lothagam (on the west side of the lake) which was over 5 million, and a left lower molar from Lukeino to the south that was maybe 6 million years in age.

No one could be sure, from these remains, just how the first prehuman was constructed. It had to be a creature of extraordinary musculature, Cann figured. Fossil femurs and other bones from later man-apes showed traction lines indicating the kind of muscular hypertrophy seen today only in athletes, slaves exhumed from the grave, or ballet dancers.

The first near-man may have borne distinct facial resemblances to the chimpanzee, since the creatures leading to both humans and chimps had been companions on the evolutionary path for quite some time (a couple million years after splitting from gorillas, if you believed in DNA hybridization), and the fact that apes had been brachiating among the branches—using their hands to swing from limb to limb, or, as was now the case with chimps and gorillas, climbing arm over arm up tree trunks after spending the day on the ground—spoke of creatures that were functioning in a vertical and semi-erect posture. An erect posture was a leap toward humanity.

If Sibley and Ahlquist were right, along with other researchers such as Masatoshi Nei from the University of Texas (whose time frame for the human-ape split generally fit that ventured by the two beleaguered Yale biologists), that meant, because of the order of descent, that knucklewalking may well have been the mode of transportation for the apes preceding Eve's earliest hominid ancestors.

But there was disagreement about that, and the only thing for sure, or nearly for sure, was that the man-apes were not tree-bound. "They were creatures that had severed the umbilical cord connecting the early hominoids to the forest," says Phillip V. Tobias, head of the department of anatomy and human biology and director of paleoanthropology at the University of the Witwatersrand Medical School in Johannesburg, South Africa.

"They lived on savannah, on the edge of the great forest. So their lifestyle was adapted to life in the open: grassland, patches of trees. In East Africa it's quite easy to see that they could have developed in the patches of savannah between river stream forest galleries. If you have a great belt of forest like the central African forest where streams come out from the forest and run into the open country, the afforestation extends like a finger down the stream from the main body of the trees in kind of spiderlike fashion. If you can imagine a little way out from the main body of the forest a couple of these fingerlike processes of good tree-covered shelter and patches of savannah between, that could well have been the scenario within which uprightness started: running against the grassy plain from one finger of forest to another. If one were upright, one would be able to spot one's enemies, like a rhino or, more likely, a leopard or saber-toothed cat."

Tobias paints the picture of a continent in crisis. Forces were at work disrupting the status quo and presumably providing the ingredients for an evolutionary upheaval. The earth was literally heaving itself up. In eastern Africa, there was an elevation of the terrain because one of the great earth plates was tilting. Lakes drained. Volcanoes spat lava. Forests were in retreat. And faunal dynamics were also shifting near the beginning of the Pliocene.

Although there are those who believe the role of environmental changes has been exaggerated, it remains an inviting theory nonetheless, something to grasp hold of for the moment, this concept of apocalyptic Pliocene disturbances. "It's a time when the drying up of the Mediterranean took place, so that over very large parts of the Mediterranean one could walk from one side to another, and there is good evidence that at that time there were major shifts of fauna between Asia and Africa," says Tobias, who was involved with important finds at a fossil site called Makapansgat in South Africa and who had assumed the professorial chair of Raymond Dart, discoverer of the Taung child.

There have been claims, from Java for example, that protohumans existed in Asia, and though proper attention has not yet been given such assertions, they remain, for now, highly questionable. "Although it would be tempting to allow one's imagina-

Australopithecus

tion free rein and suggest that hominids were one of the creatures that walked into Africa from Asia as part of that faunal shift, there is no good evidence for that whatsoever," Tobias warns. "In fact all the genetic evidence relates man most closely to the African apes, chimpanzees and gorillas. But it looks as though for the first time a number of Asian forms start appearing in the African record. And so the biotic environment is changing, the climate is changing, it's a time of major crisis within the continent, and it's against that background that from within the continent of Africa some groups of hominoids developed the upright stance and bipedal gait and this was, on present evidence, the beginning of the family hominid—*Hominidae*—the hominids."

Or, as Dart had formally named them, *Australopithecus*. It is a different genus from *Homo*, for a hominid, as the term will henceforth be used, is not even yet a primitive human. All humans were derived from hominids, according to evolutionists, but not all types of hominids evolved into humans. The most revealing *Australopithecus* was the "Lucy" skeleton discovered by Donald Johanson in eastern Africa near a great continental crack in the Ethiopian badlands. Lucy, or specimen AL 288-1, is about 3 million years old, with proportionately longer arms and shorter legs than a human possesses, a pelvis much more similar to a man's than an ape's, a dental arcade that showed signs of becoming rounded (unlike the strictly straight and nearly rectangular maxilla one finds in a chimpanzee), with canines that were no longer so fanglike and large molars but small incisors and less of a diastema, the space between teeth that accommodates an ape's clenched fangs. The hands of this species had long and curved fingers, narrow at the tips and hinting that there was still a lot of climbing in the trees. Boughs and limbs were where it may have made its bed. The thumb could do many things but it wasn't fully opposable, meaning it couldn't touch the fingertips as a human's can, and though some have suggested that flake tools in Ethiopia's Omo region were associated with man-apes—and there is good reason for thinking this, since tools which have been found in the Hadar itself, at Kada Gona, are perhaps 2.5 million years old—there is no proof yet that the earliest Afar hominids fashioned even the simplest stone implements. Lucy's skull, which was not intact, had probably been somewhat larger than a softball.

One thing was certain, and it was what most distinguished these creatures from pure apes: Lucy was not a knucklewalker but walked bipedally—on two legs and two legs alone. As Richard Leakey, once director of the National Museums of Kenya, points out, "The attainment of bipedality and upright posture probably signifies a very major change in the way these animals made a living. And it is probably at that level that you would draw your generic distinction. I think the main distinction we should be looking for is the most fundamental, which is locomotion."

The discovery of Lucy fascinated Rebecca Cann, despite her strong feelings that the much later arrival of modern humans was even more exciting. To see an artist's rendition of what Lucy and other early australopithecines looked like was to stare spellbound at a furry intermediate with wide nostrils, a bit of a forehead, eyes that hint at self-awareness, and a whole lot of thick, jutting jaw. Heavy on the eyebrows, too, of course. The skin under their coarse hair probably ranged in color from grayish-pink to black or several shades of yellow. The hair itself could have been black, brown, reddish, or silver. These man-apes were short—Lucy was under four feet in height—and yet their uprightness gave them a deceptive stature, and males were perhaps half again or twice as much in weight. For their size they were extremely thick of bone, long of arm, and, as Cann points out, exceptionally well muscled.

The brains were more than twice the size of an ancient ape like *Proconsul* but just a little larger than a modern ape like the chimpanzee, probably ranging, Don Johanson says, "from below 400 cubic centimeters to close to 500 cubic centimeters, well below what anybody would consider putting in the genus *Homo.*" The long arms of *Australopithecus* implied that there was still a good deal of climbing—again, perhaps implying a substantial amount of time, especially at night, spent in trees—but curved fingers aside, the hands were becoming quite humanlike. And use of the hands may be tied, some speculate, to growing brains.

With free hands a male hominid may have begun carrying food back home, able to roam farther and farther away, while the female could securely carry the offspring in her arms. There is not much evidence on lifestyle, but Alan Walker of Johns Hopkins says the type of teeth in the man-apes are the type that fit those of vegetarian animals. "Animals that are full-time carnivores have to cut up skin and meat for their living, and they do so with specialized cutting teeth that have sharp blades on them," notes Walker, who has studied the teeth of Australopithecus with an electron microscope. "In cats and dogs they're like scissors in the back of the mouth. Those type of teeth are entirely lacking in australopithecines."

Others who have used the same electron microscope argue, to the contrary, that pitting, flaking, and scratching on the fossil teeth indicate that later man-apes lived the life of an omnivore, just as humans do today, getting their meat not from hunting (though this may have been practiced in forms such as ganging up and cornering manageable prey), but by scavenging lion leftovers or capturing helpless baby mammals.

Still, when one pictures an Australopithecus, especially an early one like Lucy, one does not picture a scavenger so much as a man-ape picking through the brush and pulling a twig through the mouth to strip it of succulent leaves. It may have been sweet-tasting berries that made man-apes descend from the trees to begin with. As for the feet, they were no longer shaped for holding on to branches. Some argue that Australopithecus may not have walked as well as we do (as Yvette Deloison of the Musée Nationale d'Histoire Naturelle in Paris points out, study of fossil calcanei—or heel bones—shows that the feet of Australopithecus seem to point to being somewhere between those of apes and Homo, lessening its stability), while others guess the man-apes may have been able to walk with even greater ease than do humans. In summary, one might visualize Australopithecus as resembling a man below the neck but a chimp above it. "We're looking at a creature that was more of a chimpanzee on two legs than anything else," says Berkeley paleoanthropologist Tim White, a colleague of Sarich's.

At around the same time that another much closer colleague of White's, Donald Johanson, who runs an independent institute in Berkeley, was discovering Lucy and a slew of other australopithecine fossils in the Afar region of Ethiopia, more than 1,000 miles to the south Mary Leakey, the grande dame of both archeology and paleoanthropology and wife of the now deceased and world-famous Louis, was heading a team that was finding the single most striking evidence of hominid bipedality imaginable: actual footprints fossilized into hardened volcanic ash. She was not only on the man-ape trail but was making plaster casts of it.

Mary's discovery was at a snake-infested and remarkably remote stretch of upland savannah called Laetoli, in northern Tanzania, about 100 miles east of Lake Victoria (and about 25 miles from the fossil haven that had made her and Louis known around the world, the red, gray, and buff-layered gully known as Olduvai Gorge). Olduvai is part of the Great Rift Valley that extends all the way up to Israel, a seam of continent tearing apart and exposing past secrets. Through the millennia it has allowed whatever was burbling beneath to express itself in a series of volcanoes. In the case of Olduvai (which originates to the west in lakes Maser and Ndutu and flows 46 kilometers into what is known as the Olbalbal Depression, alongside the bleached Serengeti Plains), the ravine was cut by seasonal rivers and was one of those lakes that eventually disappeared after swamping old bones with preserving sediment. Hominids tended to congregate at just such lakesides, interested in staying within a few hours of water. When they died, their bones, if untouched by scavengers and left where the lake expanded or a seasonal river drained, would be covered with sediment that eventually turned into rock, the bone itself transforming partly into stone as minerals replaced the calcium.

Ash from the volcanoes has also preserved fossils, allowing them to be dated with precision. At Olduvai such ash has formed strata as distinct as they are beautiful. At one time this area was a lakeside where hominids began using the most rudimentary of implements—flakes knocked off large volcanic cobbles to make scrapers and choppers—and it was those tools, along with the prospect of finding the bones of the hominids which had used them, that had brought the Leakeys to Olduvai.

Any search for Eve inevitably leads back to the Leakeys.

The son of Kenyan missionaries, able to speak fluent Kikuyu and inducted into one of the tribes, Louis first began exploring 300-feet-deep Olduvai in 1931. By the time he met Mary a couple years later, at a dinner following a lecture Louis was giving in England, he was already touting spectacular fossil finds such as

"Oldoway Man," a modern-looking skeleton that at the time seemed a million years old, and skull fragments from a place called Kanjera, along with part of a jaw from a nearby site known as Kanam, which Leakey took to represent a new species, *Homo kanamensis.*

It was not the most auspicious beginning: Oldoway Man, which had actually been discovered by a German scientist, was thought by critics to be much more recent than Leakey claimed (an intrusive burial into older sediments, granting it a falsely primitive age), and the Kanjera skull fragments, some said, were also quite young, perhaps only 15,000 years old, for now frustrating Louis's goal of proving that the *Homo* lineage went way back in time. Leakey didn't believe australopithecines were in our lineage. Instead, he believed the precursors of Eve (or "Adam's ancestors," as he more masculinely phrased it) were primitive humans that had co-existed with the man-apes. He was also determined to show that humans originated not in Asia, as most everyone else had believed after the discovery of Peking Man, but instead on his home turf of Africa.

Contrary to common perception, it was his wife Mary who actually made the most significant of their fossil discoveries. The daughter of a painter, with a keen interest in prehistoric art (and, on her mother's side, a direct descendant of John Frere, famous in the literature of British archeology for his recognition of stone tools, including "hand axes" at a well-known site called Hoxne, and the first person on record as claiming a great antiquity for man), Mary had a good measure of colonial blue blood in her veins, one of her Frere ancestors a baronet, the family so well known in the fight to free slaves that today (in India, South Africa, and Kenya) are several locales with the same name: "Freretown." She attended geology lectures at London University and worked at Swanscombe, in England, where a famous skull would soon be found. Later she would often find herself photographed with a cigarette dangling from her lips or even puffing on a cigar, but she was a sophisticated young lady who looked like she had walked out of the pages of a Hemingway novel. At the age of twenty-two, she fell in love with Louis and joined him at Olduvai in 1935. Although she found hominid fragments right away, most of their early work would be categorizing stone tools and copying cave art. Nearly a quarter of a century would pass before there was another major hominid find, in 1959, a skull which Mary found and which Louis named *Zinjanthropus*, or *"Zinj"* for short. It was initially thought to be the Olduvai Gorge's toolmaker even though it was not *Homo* but quite clearly an *Australopithecus.*

So massive were *Zinj*'s jaws that it quickly picked up the nickname "Nutcracker Man" (though the Leakeys preferred to call this fossil "Dear Boy").

Louis was himself a robust and Runyonesque character, with fierce powers of concentration and polemic, who couldn't pass up an opportunity for drama and flair. When Clark Howell, as a young field researcher, stopped by the Leakeys' for dinner, Louis shocked him by bringing out a biscuit tin for dessert and opening it to reveal the incredible *Zinj* fossil.

Louis soon helped Howell begin an exploration of the fossil-rich Omo Valley. Little could Leakey have known that the city where Howell would eventually plant his feet, Berkeley, would later serve as the home base not only for the new-fangled geneticists who would create the Eve hypothesis, but also for a coterie of paleoanthropologists who would attack his family with the same fervor as Sarich and Wilson attacked Sibley and Ahlquist.

But that's getting ahead of the story. What is important just now is that *Zinj* the man-ape galvanized scholars everywhere and overnight made paleoanthropology fashionable. Scientists the world over took a new interest in the evolution of humans. When Olduvai Gorge was dated with a technique of geochronology, the oldest bed, Bed I, where *Zinj* was discovered, appeared to be 1.8 million years old. This forced paleontologists to double the age for the beginning of the next epoch, the Pleistocene, which is now considered to have begun about 1.7 million years ago.

But neither *Zinj* nor any of their subsequent discoveries would titillate Mary as much as what she found, in the 1970s, on the thorn-bush plains of Laetoli. She and Louis had visited there in 1935, summoned by a Masai native who told them about "bones like stone" there, but most of their work continued to center on Olduvai. When Laetoli was finally explored in a system-atic way a few years after Louis's death, it yielded dozens of homi-

nid teeth and a mandible very similar to what was found in the
Afar but even more ancient—perhaps 3.5 million years old, which
threw it back into the Pliocene, probably the oldest good hominid
samples ever found. The "Leakey luck" had come surging back to
predate Johanson's Lucy.

In 1976, members of Mary Leakey's team happened upon
a stretch of hardened volcanic ash containing the extraordinary
fossilized tracks of ancient insects, birds, hares, baboons, dik-dik
(a tiny antelope), rhino, giraffe, ostrich, a saber-toothed tiger, pigs,
and an extinct relative of the elephant at Laetoli, a savannah where
Masai warriors still carried spears and giraffes moved between
distant acacias that were shaped rather like weary umbrellas. These
fossilized hoof marks were an incredible snapshot of wildlife movement perhaps as long as 3.7 million years ago, but in 1977 Leakey
was able to confirm something even more amazing: among the
animal trails fossilized in the volcanic ash were also the footprints
of three bona fide hominids.

There were sixty-nine tracks, hardened in the tuff like ce-
ment and belonging, it was surmised, to at least one male and
perhaps a female with a youngster skipping alongside. There it was,
etched in stone, a dramatic confirmation that hominids (as Lucy's
pelvis also so strongly showed) had been walking much like we do,
perhaps with a more shuffling gait, nearly 4 million years ago. The
arches and balls of the feet were much like those of modern
humans. Other paleoanthropologists could collect foot bones if
they wanted. Mary Leakey in her waning years was going to come
up with footprints.

We must always keep in mind how very long ago this is:
these hominid prints were about 700 times older than the Pyramids at Gîzeh. The man-apes whose feet had impressed themselves
so indelibly into the rain-moistened volcanic ash appeared to have
been somewhat bigger than Lucy, perhaps more than four feet in
height, and while there has been speculation that they were fleeing
an eruption from the volcano Sadiman, which is about twenty
miles away, Mary told me there was "no hurry at all. Not with the
hominids or animals. Everything was walking quite calmly and
sedately. As far as we can make out it was a completely modern
gait, but slow, short strides. The weight distribution from heel to

side of the foot and toes was the same as ourselves."

Following the path also caused Mary to retrace the hominids' very frame of mind. There seemed to be a point, for instance, when one of the travelers stopped, paused, and glanced around looking for a possible threat or anything out of the ordinary. "This motion," she has said, "so intensely human, transcends time. Three million years ago, a remote ancestor—just as you or I—experienced a moment of doubt."

The footprints were uncovered and cast into plaster with the help of budding paleoanthropologist Tim White, who tirelessly chipped away at the gray ash with dental picks, beside himself with astonishment at these eerie signatures from the past. This was the real stuff, a far cry from cleaving DNA molecules. "Each morning I would literally run out to the site to keep digging—to see what was going to be under the next little patch that was going to be removed that day," recalls White. Mary observed that there appeared to have been no campsites or home bases like those at the base of the Olduvai section, causing her to assume that these hominids, twice the age of *Zinj,* lived in roaming bands like today's chimps and baboons, scavenging and gathering food as they moved from spot to spot. At one point the trail had been crossed by a three-toed horse known as *Hipparion.*

If White, whose volume-lined office is in that cavernous place at Berkeley called Kroeber Hall, was originally a close friend of the Leakeys, this very soon would dramatically change, for he was teaming up with the Leakeys' new rival, Donald Johanson, whose discovery of Lucy made him the only paleoanthropologist who could hope to challenge the Leakeys' domination of the field. Much to Mary's distaste (and also that of her son Richard, who was quickly replacing his father as Africa's—and the world's—most famous fossil hunter), Johanson and White, upstarts from the United States, claimed that Mary's Laetoli hominids were identical to the man-apes Johanson had discovered in the Afar. They had even designated them as the new species called *Australopithecus afarensis,* in recognition of the region where Lucy and accompanying fossils had been found.

Mary was livid. She didn't believe her Laetoli fossils were man-apes. In keeping with the Leakey line, she saw them as ancient

Homo. She detested the name *Australopithecus.* While Lucy was clearly a man-ape, she and Richard, who had visited Johanson's camp a day before Lucy was found (as if allowing some of their luck to rub off on him), thought some of the hominids Johanson's team had unearthed, which resembled those at Laetoli, were not *Australopithecus* but human. It was as if Johanson was not just trying to upstage her, but was virtually confiscating her fossils. He talked about *her* findings at a Nobel Symposium in 1978 before she had a chance to speak. Worse, Johanson and Berkeley's White decided that the "halotype" or type specimen used to define their new species *afarensis* would be a mandible discovered in 1974 not in the Afar but at Mary's own Laetoli. She thought it wrong to lump them all together, and naming a fossil found in Tanzania after the Afar (which is in Ethiopia) especially irked her.

"It's a very complicated situation, but there were two species at the Afar, not one, and I think that the Laetoli hominids have resemblances with the bigger of the two at Afar," Mary told me, when we met in the café at Hemingway's old haunt, Nairobi's Norfolk Hotel. "But part of the quarrel—or whatever you want to call it—was the use of the Laetoli mandible, which was in Tanzania, as a type specimen for *afarensis,* which is named after the Afar. And this is just not found in sound scientific circles. It is something to which I objected quite strongly and still do. Geographically, it's not acceptable. It's a total misunderstanding of the material. [White and Johanson] didn't take the time or trouble to study it thoroughly before they rushed into print with '*Australopithecus afarensis.*' It was a rush job because they were frightened that someone else might name it."

Others, such as Harvard's venerable Ernst Mayr (whom Johanson has described, ironically, as "the last court of appeal on the scientific naming of things"), is also upset, because in naming Lucy and the Laetoli hominids *Australopithecus afarensis,* the renegades from Berkeley were pushing aside Dart's original name for the australopithecines, *africanus,* which until then had been thought to serve as man's main rootstock. "That was the most idiotic thing, it just shows that Johanson doesn't know what it's all about," snorts Mayr. "*Africanus* and *afarensis* quite likely were geographical races of the same species!"

But Johanson saw a number of differences between the australopithecines found in South Africa, where *africanus* was chiefly located, and those further north. For instance, the Ethiopian specimens had a more primitive pre-molar. And whatever the case, the new name *afarensis* has become the official designation for the oldest known hominids, followed in time by the *africanus* species.

Issues like this—and there is no better example than australopithecines—are often entirely muddled by nomenclature. At one point back when Mary Leakey was making her first major discoveries in the 1950s, there were assigned to man's various and alleged ancestors at least twenty-nine different generic names and more than a hundred species names—"a totally bewildering diversity of types," noted Mayr. Of course, many species with different names were really the same thing. The history of this peculiar science is filled with instances of dubious new names being assigned to fossils because of confusion caused by the fragmentary nature of fossils or because political purposes—such as promoting one's own discovery as especially new and important by issuing it a new name—outweigh scientific purposes. As Alan Mann is fond of cautioning reporters, "in this field a person kicks over a stone in Africa, and we have to rewrite the textbooks."

Richard Leakey, jibed by Johanson for implying, as his father had, that australopithecines were not man's ancestors, says the taxonomic scheme being used to describe and discuss the fossil record of these creatures that were basically chimps on two legs "is totally inadequate to actually indicate and signpost what has probably happened. The current understanding from molecular biology, DNA, et cetera, is that our comparability to chimps is such that a family distinction is probably exaggerated. Clearly, at some point in time, you will go back beyond the point where there is a large break. Is that something that is going to be called *Australopithecus* or is it something that will be called something else?"

There is, of course, no answer to what immediately preceded *afarensis*, but we do know that the Afar and Laetoli hominids appear to have been followed by the slightly younger *africanus* specimens from South Africa and similar specimens from the Omo River region in southern Ethiopia. *Africanus* lived 1 to 3 million years ago, was about the same size as *afarensis* (three to five feet, depending on whether it was male or female), and is represented not only by Dart's Taung child but also by discoveries made by Robert Broom (an eccentric physician who used to show up at quarries in the traditional doctor's garb of black suit and starched wing collar looking for endocranial casts amid the limestone breccia and hoping to nab any valuable fragments before they were doomed to a quarry's kiln burner. If there were no fossils immediately available from mining operations, Broom was known to plant dynamite in the breccia—that mix of sand, stone, and limestone that so often encrusts a fossil—and blast the fossils out).

There are so few differences between *afarensis* and *africanus* that until a few years ago Phillip Tobias, official keeper of *africanus* specimens, refused, like Leakey and Mayr, to acknowledge the new species Johanson had named. "There are clearly a great number of very close resemblances between *africanus* and what Johanson and his colleagues have called '*afarensis*,' and I was on record as saying *afarensis* was nothing more than the East African equivalent of *africanus*," Tobias explains. "The two forms are very close. The pelvic bone of Lucy and pelvic bones from Sterkfontein in South Africa are virtually indistinguishable."

The closer one got to the rootstock of humans, the more controversial the issues. But Tobias, during the past eight years, has come to recognize the difference between these two man-apes, as have most paleoanthropologists. *Afarensis*, he says, is indeed more primitive, with a slightly smaller brain size than the South African australopithecines, and hand bones that are more chimpanzoid than what was found at the South African cave called Sterkfontein.

The important point is that mankind appears to have been preceded, before any Garden of Eden, by creatures more peculiar than anything in the history of zoology, certainly creatures more wondrous than *Aegyptopithecus* or any of those first tiny Egyptian apes. If current chronologies are at all reliable (and new fossils may one day render all current conceptions obsolete), a third species of *Australopithecus* arose around *africanus*'s time, this one, however, with a bigger, flatter, dish-shaped face, flaring cheeks, and a far heavier jaw than either *africanus* or *afarensis*.

Discovered first by the Leakeys at Olduvai (in the form of the fossil *Zinj*), the big-jawed type of hominid was named *Australopithecus boisei* after a businessman in England, Charles Boise, who had funded the Leakeys' expeditions. Another example of *boisei* was found by son Richard on his first independent venture: an expedition in 1968 to the shores of Lake Turkana. It was also at Turkana, in northwestern Kenya, that Alan Walker of Johns Hopkins, one of Leakey's closest colleagues, discovered a hominid form that may be the intermediate linking an earlier australopithecine like *afarensis* to the apparently later and larger *boisei*. It is known as the "black skull," or *aethiopicus*. Discovered in 1985 and heralded as the most important find since Lucy, it has been dated to 2.6 million years old.

It is probably safe to say, according to Frederick E. Grine, an anatomist and anthropologist at the State University of New York at Stony Brook, that *boisei* and similar types were bigger in body size than *africanus* and *afarensis*, perhaps even as tall as five feet at the top end of the scale (and something on the order of four feet at the low female end), justifying those big jaws. It also had a brain about 50 cubic centimeters larger than *afarensis*. Unlike those earlier hominids, it possessed a strange, striking ridge of bone on top of the skull, which rose from the low, rugged cranium front to back like a cutting blade. This is the "sagittal crest," which is also seen in apes, and it anchored the muscles powering those formidable jaws. A visorlike crest was formed over the eyes, and inside the skull, meanwhile, were extraordinarily large, grinding molars, the better to chomp down large volumes of leaves or fruit—the same diet perhaps as an orangutan or chimpanzee.

And unlike earlier man-apes, it seemed to be totally terrestrial, spending little or no time at all in trees. These larger, more robust australopithecines were, according to Randall Susman, another professor at Stony Brook who has extensively studied primitive limb bones, essentially full-time terrestrial bipeds.

Boisei is thought to have eventually gone extinct, as has another robust species of *Australopithecus* which Broom found at a South African site called Kromdraai many years before the *boisei* finds. This species (also discovered at yet another site, called Swartkrans, just across the valley from Sterkfontein), was not as massive in the face as *boisei* but nonetheless possessed a heavily buttressed jaw, warranting the name *robustus,* and leading to speculation that its diet was like a monkey's—fruit and leaves but also some tough stuff like nuts and possibly even some bark. *Robustus* was like *boisei* in many ways, including body size, and it had a thumb that was fully opposable, strikingly similar to a modern man's. Some of the foot bones and indeed parts of *robustus*'s big toe were a spitting image of a modern human's. The gait cycle was very similar to man's, according to Susman, although *robustus* didn't lever off its great toe like we do, but instead levered more off the outside of the foot.

The Leakeys themselves at first thought some of the stone tools at Olduvai belonged to a robust *Australopithecus* like *Zinj,* and it is now known that these bigger australopithecines had a fairly precise hand grip, giving them the ability, perhaps, to take a tennis-ball-sized cobble and knock flakes off it to create the scraping and chopping implements Mary had so assiduously catalogued at Olduvai. The thumb had been getting increasingly "opposable" since the later man-apes and was now virtually human-like. Even back as far as Lucy or *africanus,* wooden tools may have been in use. Chimpanzees, with less brain size than the most ancient *Australopithecus,* are known to throw stones at baboons, to warn intruders away by wielding large branches, to use leaves as a sponge, and to strip sticks and employ these to extract edible termites from the sandcastlelike nests.

Were the later man-apes the first to fashion specific tools? Tobias didn't think *Australopithecus* had gone any further than crude wood implements, but current investigators criticize that viewpoint as old-fashioned. "If we look at the fossil record of human evolution as we know it today, the earliest tools are dated at about 2.4 million years at Lake Turkana," argues Randall Susman. "And they're found in deposits that only contain robust australopithecines—*boisei.* The first *Homo* appears in the fossil record 400,000 years later, at two million years ago. Some people say, well, there must have been *Homo* around, we just haven't found it yet, but at Swartkrans, dated from that cave at 1.8 million, we have stone and bone tools and the predominant hominid at that time was *robustus,* so again, in South Africa, there's evidence that some of the earliest tools were found in association with robust australopithecines."

From what we can surmise, these predecessors of Eve were more sophisticated than the chimpanzee, the males roaming farther, the females collecting fruit near the home base and perhaps carrying it home, grinding seeds, grass, hard-skinned fruit, and gritty roots with their huge molars (although diet probably varied by species and by habitat, with some staying away from roots and grass), perhaps scavenging a little meat from a lame gazelle or what a lion left behind (although, as we've seen, many claim the australopithecines were all vegetarians), tasting bird eggs and frogs and lizards, maybe even swallowing mice whole, grunting with satisfaction from time to time or yelping like an ape, for they could not talk, or jumping up and down, madly thumping the ground like a comic chimp.

There were probably a dozen or more man-apes to a group, constituting what was in essence an extended family. Perhaps they were monogamous, with the beginnings of marriage, or what the scientists so unromantically refer to as "pair bonding." But gorillas and chimps aren't monogamous, having "reproductive access" to just about any female in a troop, and so perhaps australopithecines followed suit. Savannah dwellers to the north, in South Africa they either lived in caves, sought shelter in them during seasonal downpours, or were dragged or dropped in them by saber-toothed cats, for that's where the bones are found, with teeth indicating an average age of only twenty-two. Once thought by Dart to be "killer apes," there is little reason left to consider them homicidal.

And so the hominids, perhaps starting with *afarensis*, branching into robust side groups on the one hand and a lineage that aimed itself toward mankind on the other, were, if you believed most paleoanthropologists, about to be thrust from the lowly category of man-apes into the classification of primitive humans.

Man-ape was becoming ape-man.

Cann, who had watched the Johanson-Leakey dispute from the close vantage point of Berkeley before heading for Hawaii, kept her eyes less on the controversies than on the bottom line. "Although anthropologists do not agree on the details of australopithecine evolution (or of any other stage of hominid evolution, for that matter)," she noted, "there is general consent that, about three million years ago, Lucy's kind began to diverge into two lines of descent, one spawning other australopithecine species—*aethiopicus, robustus* and *boisei,* which ultimately would become extinct—and the other giving rise, a million years later, to the genus *Homo.*"

11

THE FIRST "HUMAN" was very possibly *habilis,* another discovery by the Leakeys, again at Olduvai Gorge. They'd begun unearthing it within two years of the *Zinj* discovery, just 100 yards away, this time two individuals, one a juvenile represented by parts of a skull and the small bones making up the hand, the other an adult, in all likelihood a female who left behind some revealing foot bones. They were contemporaries of *Australopithecus,* at least as old as some of the man-apes, but different in certain crucial regards, with a skull that was not nearly as robust and flaring as a *boisei,* not as chimplike as *africanus* or Lucy, but instead bearing some surprising resemblances to man.

    Habilis was really more humanoid than human, as close as anyone has yet seen to the dividing line, the intermediate stage, between man-apes and *Homo sapiens.* Mary Leakey continued to dig out additional fragments of this missing link, more pieces of the puzzle fitting into place. *Habilis* appeared to be the one associated with the stone tools at Olduvai (as opposed to *Zinj*) and it looked as though it had been derived more from the slighter, gracile australopithecines like *africanus* and *afarensis* than those robust nutcracker types that are believed now to have gone extinct. Its molars were smaller in proportion to its incisors, and its brain, upon evaluation by anatomist Phillip Tobias, appeared to be just under 700 cubic centimeters, a big leap from the 500 cubic centimeters of the average robust *Australopithecus. Habilis* had lived

Homo habilis

on the order of 2 million B.C., nearly halfway between the first man-apes and modern humans.

Louis, ever in search of ancient "true man" (as opposed to the australopithecines, which he called "near man"), immediately decided the new Olduvai discovery was indeed an ancient species of the *Homo* that he had sought for so long. "Although Louis Leakey was convinced from almost the beginning that this was something to be put in the genus *Homo*, I was the one given the specimens to describe by him and Mary, and I was not convinced, and for four years I sat on these specimens, studying them, comparing them, measuring them, comparing them with everything else available from South and East Africa, from Java, and so on," recalls Tobias. "And it was only when a number of further finds came to light in 1963 and I saw the departures from *africanus*-like structure that I came to realize that we had here a new and different population—not just a freakish individual which was a bigger-brained member of *africanus*."

Bigger brain, smaller teeth. Ape-man. "*Habilis* retained the short physique of his forebears, but had a less protruding face and a higher, rounder, and significantly larger braincase," explained Cann.

When Tobias considered the geological levels at which this humanoid was being found at Olduvai, especially "the extra evidence pointing to this chap having been a toolmaker," he decided it was indeed *Homo*, "the most primitive species of *Homo*," as opposed to simply a highly advanced *Australopithecus*.

Whether or not early man-apes were using the crudest sorts of wood implements (which, because they are perishable, do not make it into the fossil record), and whether or not the more recent man-apes like *boisei* (which existed alongside *habilis* for hundreds of thousands of years) had taken the giant step to actually fashioning pieces of stone, perhaps to scrape bark off trees in search of grubs, *habilis* remains the only one definitively linked to the watershed event of a stone tool industry. No one doubts this evolutionary link was involved in altering big pebbles to cause a cutting edge, and perhaps doing other sophisticated things as well. It was the onset of the Stone Age. You took a big cobble of lava, smashed it against another rock, and when the flakes flew off the side they made scrapers, while the core was a chopper, with a sharpened edge.

"What to call it?" says Tobias. "Well, there was one man around who had a facility with the English language, a kind of genius for inventing new names, and that was Raymond Dart. Dart had invented that absurd name 'Australopithecus'! So I went to Raymond and said, 'Raymond, we've got what we're convinced is a new species, we're convinced it is the first confirmed manufacturer of tools, we're convinced it belongs to the genus Homo, it shows the beginning of the brain enlargement well beyond the ape size that characterized Australopithecus, it's nearly 50 percent bigger in brain size, it's got these reduced teeth, and it is well on the way to being us. So we want to call it Homo but Homo what?' And Raymond went away and thought about it and came back with the simplest, the shortest, the sweetest name that he'd ever invented, and that was 'habilis.' "

Which meant handy, skilled, dexterous. Homo habilis. Reporters called it—him—"Handy Man." Critics called it ridiculous. There was another of those long storms of protest, lasting now more than twenty years. Different as the skull was, it still was not nearly anything that one could confuse with a modern cranium, and at four and a half to five feet (or perhaps even as elfish as Lucy), this little ape-man, in artists' reconstructions, still had a chimplike face and all that coarse hair. In many ways, upon first glance (and second glance too), it could have been confused with the gracile man-apes like africanus—which is what the critics thought it was and what a good number of paleoanthropologists still do.

The controversy lingers, but habilis has never been dethroned as the first human. "Leakey luck" indeed: Mary, discoverer of that ancient ape Proconsul, discoverer of rudimentary stone choppers, which became known as the Oldowan culture, discoverer of the first boisei, now, with her husband and another son, Jonathan, had unearthed a specimen to help fill that wide chasm between man-apes and Homo sapiens. The announcement of its name was made in Nature in 1964.

Besides habilis's exceedingly primitive appearance (which again causes anyone taking a first look at these skulls to pause and wonder a bit how it could be considered human), there was the matter of brain size, which fell just below the 700 to 800 cubic centimeters accepted as the threshold for humans. (The average for modern humans is 1,300 cubic centimeters, nearly twice that of the average *Homo habilis.*)

Ralph Holloway, a biologist at Columbia University, has made latex endocasts of fossil skulls with telltale dimpling on the inside. That is, the brain imprints itself into the bone during the lifetime of an animal, and its patterns vary between species. Besides the differences in size, he can tell a *habilis* from the man-apes by the shape and proportions of the various brain regions, which, in *habilis*, are increasingly more humanlike than apelike (the frontal lobes seem to increase during the course of evolution).

"One thing you see is a very striking asymmetry between the left and right sides, so that the left occipital portion juts out further behind than the right side and the right frontal lobe is wider than the left as opposed to what you see in australopithecines, where you find little difference," says Holloway. "This is a striking difference, and that's associated with handedness."

Handy Man. And if it had tools, *habilis* was probably cutting meat with them. Perhaps the australopithecines were just vegetarians after all. Randall Susman, however, says that the man-apes that preceded *habilis* also may have been using tools for culinary purposes, but with a different menu in mind. "I think what happened is that *robustus* was using tools for basically a vegetarian diet—getting plant food and processing it, getting inside hard roots and opening shells—and probably *Homo* was doing that as well but also eating more meat than the robusts," he says. "The microscopic patterns on the teeth, the wear patterns, show that that's probably true. You get a much more carnivorelike pattern of wear on the cheek teeth of *Homo* than robust australopithecines, whose teeth look like part of a mortar pestle that is used for grinding things."

There are those who have hinted, without enough proof to quite back it up, that *habilis* may have been the first to use fire, or if not, may at least have begun to understand the effects of lightning striking trees and how flames could spread from a good thunderstorm amid the trees or from molten lava descending from the highlands.

It is tempting to imagine *habilis* occasionally bringing a burning branch back to camp and experimenting with the ethereal orange flames dancing from the fire. Was cooking about to begin? No one can really say. The first ironclad proof for intentional use of fire comes a bit later. Right now the important culinary point is use of meat. Many anthropologists have speculated that meat eating may have been what spawned the first close-knit social structures, for where vegetarians are prone to go off by themselves and munch along the edge of a forest, meat eaters would gather together and share a carcass dragged back to the home base or a seasonal camp. The key word here is "sharing." It implies increasing dependency on group functions and perhaps even the glimmerings of a barter system: you find the water hole, I'll find a carcass. And bring some berries back in that contraption we wove with leaves and branches.

This was probably still said, of course, in crude grunts and pointed gestures, if it was communicated at all, if there were as yet anything like woven baskets. Excepting certain developed instincts and skills, the first human may have had less cognizance, for all we know, than a kindergartner. But there are those who believe that *habilis*—whose bones have been found from Sterkfontein in South Africa to the Omo region of Ethiopia, often alongside australopithecines of the robust and *africanus*-like types—was capable of language. There appears to have been an enlargement (or at least greater complexity) in the Broca's area of *habilis*'s brain, which governs the motor functions of speech. These advances appeared, interestingly enough, at roughly the same time as stone implements.

Despite popular notions that only one kind of proto-*sapiens* was alive at any particular period, *habilis* was hardly by himself, with all those man-apes scurrying through the terrain: robusts in the south and the super-robusts, or *boisei,* in the east, along perhaps with *africanus*-like leftovers. There were so many hominids around, in fact, that the *habilis* species has been criticized as simply being a varied grab bag of specimens that bore characteristics of other species but didn't quite fit all existing definitions and so were lumped into a convenient, catchall species designated to *Homo.* The same criticism has been applied to the classification of *africanus.* Though their bones are often found together, this doesn't mean *habilis* and the man-apes necessarily shared the same ecological niche. To the contrary, they may have used the same territory at different times of the year or taken food at slightly different levels, just as wildebeest and zebra often migrate to grazing areas a few weeks apart. They may well have moved about, in other words, on a seasonal basis, depending upon where herds were moving or where the vegetation they enjoyed was in bloom. Perhaps they mixed with man-apes just as chimps sometimes mix with baboons.

Richard Leakey points out that if three types of hominids co-existed, their interactions might have been like various species of monkeys that overlap in niche utilization during times of plenty but revert to their particular specializations in times of ecological stress. The meat eater might have eaten more meat during dry weather, the tree climber might have climbed more trees in search of fruit beyond the vegetation normally gathered on the ground, and during drought the one with the most powerful jaws might have increased its portions of foods that the others weren't as adapted to eating.

Besides scavenging, Mary Leakey has seen evidence at Olduvai of successful hunting. Fractures over the right orbits of antelope skulls indicate blows delivered at close range, perhaps with a club. There is even one site that may have been used for butchering an elephant, which perhaps had become trapped in a swamp or had been purposely driven there (though this would seem more in accordance with the practices of later, more evolved ape-men). In one layer at Olduvai, lower Bed I, there was what appeared to be an artificial structure consisting, says Mary, of a circle of loosely piled rocks resembling the foundations of the kind of crude shelters of branches and grass made by current-day nomads. But this circle of stones is nearly 2 million years in age! In other words, temporary shelters of the most rudimentary sort may have been under construction in the days of Bed I. She guesses a group of habilines may have been a dozen or so in number.

So a picture continues to form, and Randall Susman sees *habilis* as "an animal that made and used stone tools, that ran around in a savannah-woodland-forest mosaic, where there was forest around the rivers and lakes, but by and large the inhabitant of a drier environment, and one who I think, unlike the robusts, who were too big, was still a tree climber and probably slept up there, to avoid predators and also climbed to get at any succulent fruit. Its limb proportions still looked somewhat like a chimp's, with huge arms. The conclusion's inescapable that an animal with large upper limbs can only be doing one thing: it's got to be using them for climbing trees. I don't think there's any evidence of weapons, but if you have a stone in your hand and a big snake comes by, I'm sure you wouldn't hesitate to throw it."

Susman directs a field study of pygmy chimps in Zaire and watches for such things. Although he says there is no way of knowing if that first "human" was monogamous, a few hints might be available from what is known as "sexual dimorphism." If a species is dimorphic, the males (as with certain man-apes) are substantially larger than the females, and when that's the case with savannah primates, they tend to live in multi-male groups where the males watch over a bunch of females with their offspring. On the other hand, "monomorphic" primates like the marmoset and gibbon, which have males and females of roughly equal size, tend to live in more monogamous, pair-bonded groups. Most man-apes were extremely dimorphic, and *habilis* might have been highly dimorphic as well.

Summing up the image of the first known humanoid, Susman adds: "I think if you saw *habilis* walking down the street, you'd start seeing an animal that looked like a small human but with bigger upper limbs."

It might decide to scurry up a telephone pole, if a German shepherd happened by, or pick up a stone in self-defense, and if it walked into a zoo, the odor of monkeys wouldn't much bother "him." If all that was seen was *habilis*'s face, the keeper might even think that one of the apes had gotten out of its cage. But its eyes would flit around more knowingly, and it wouldn't go off dragging itself behind its knuckles, for its two-legged gait was fully upright and (like *robustus*) it had fairly modern feet.

Consider all the preceding information subject to debate and we can go from the hypothetical images of *habilis* back to the fossil discoveries themselves. In the case of *habilis*, this meant not just another inevitable argument but a situation in which the most famous paleoanthropologists declared war on each other. At issue was the best preserved *habilis* skull, "1470," found by Richard Leakey at Turkana in 1972. Before the discovery of Lucy, and then the footprints at Laetoli, the unearthing and preparation of this famous skull had portended to be the most dramatic paleoanthropological recovery of the decade. It was the fossil that made Richard famous, as *Zinj* had made his parents famous.

Born in 1944, Richard E. Leakey had at times strayed from his parents' line of work, determined to set out on his own and so filling his early years with work as a safari guide. Richard didn't want to live in his father's shadow. There was much more in Africa—watching the incredible wildlife, which would be a lifelong passion, for instance, or following Masai on a terrifying lion kill—that was his for the taking. Though of British descent—the "white hunter" in a land of mud huts and fifty tribes—he was fiercely proud to be a Kenyan. As a kid he had spoken Swahili better than he'd spoken English. School held little allure for Richard, and he'd had enough of it by age sixteen. The savannah, the bush, the veldt of eastern Africa were all his classroom.

It had been no standard childhood, not with Louis as his father, leading the kids to Lake Victoria to look for Miocene apes and firing a gun into crocodile waters to clear the way for a needed bath. There were the khaki-drill trousers and dusty, bouncy roads, adventure everywhere a boy could turn. How many kids go looking for rhino? In his youth, Richard once set about trapping lions that were haunting domestic animals near the Leakeys' home in the Langata suburb of Nairobi. In later years his criticism of the government's inability to halt elephant slaughter would temporarily cost him his job at Kenya's national museum. It was the museum where his father was curator, and he was just a kid when he was making plaster casts of fish. To young Richard, a fossil bone, with obvious exceptions, was no big deal. It was like a stock report to a child of Wall Street. And talk of geological tuffs came with dessert.

So it would be no great surprise if he was bored at times by the thousands of stone tools his mother had collected (looking at first glance like nothing more than chipped rock) and the interminable discussion of this kind of femur and that kind of mandible, or the wrinkles in *boisei* molars. There was only so long that you could stare at an ape tooth or a sliver of sharp-edged rock. But prospecting was in his blood, he never resisted paleoanthropological work for very long, and in his wanderings he couldn't help but notice potential fossil sites.

It was a good thing, too, because Richard was endowed with more "Leakey luck" than perhaps anyone else in his family. He found his first fossil at the age of six (the jaw of an extinct pig), and at the age of nineteen, while piloting a light plane, he had spotted sediments near Lake Natron that soon yielded the first lower jaw of a *boisei*. What took his parents years to find took Richard weeks. He had spotted the Turkana site during another flight, this time winging back from an expedition in the Omo, and within three weeks of his first mission to Turkana he found another *boisei* jaw. In 1969, the face of a super-robust *Australopithecus* stared up at him from a dry streambed, just waiting to be plucked up and carried to the museum.

Other fossil hunters like Clark Howell could spend forever at the Omo without coming back with a famous skull, but for Richard it was as if a third eye led him to just the right place. Turkana was a gold mine. It spewed fossils like a fountain: fossils representing about 110 hominids within five years, dozens more by the end of the first decade of exploration, so many that in one paper alone, presented in the *American Journal of Physical Anthropology*, Leakey reported 58 new fossil hominids—comparable to a fly fisherman coming back from an afternoon with a truck full of trout. Among the Turkana finds was one so odd—"1813"—that it could not be and to this day is not categorized with any particular species but is rather considered either a small, female *habilis* or some other kind of previously unknown and extremely ancient *Homo*, with teeth just like *habilis* but a strikingly smaller cranium.

None of these fossils, however, grabbed the same attention as 1470, a *habilis* skull that was complete, save for the base and lower jaw. When it made the cover of *Time*, along with a story starring Richard, the issue was one of the magazine's all-time best-sellers, selling about as many copies as covers devoted to the embattled Richard Nixon during Watergate and even more copies than a sexy issue on Cheryl Tiegs.

Clearly, Richard was a superstar.

Pieced together by Richard's wife Meave, a laborious process that consumed six weeks because of the fragmentary nature of the parts, 1470 presented a nearly complete version of a *habilis* face; the brain was determined by Richard's team to be 775 cubic centimeters.

The overall shape and thickness of the cranial vault was similar to *africanus*, noted Bernard Wood of the Middlesex Hospital Medical School in London, but one major difference was the larger size. Although the females were small, some of the habilines, it appeared, may have been similar in size to the robust australopithecines. It was certainly hard to see a skull as big as 1470 placed atop the body of the smallest man-apes.

"This was fantastic new information," said Richard, for they now had "an early fossil human skull with a brain size considerably larger than anything that had been found before of similar antiquity."

Indeed, that was just about the entire issue: antiquity. The *habilis* skull from Turkana, a site that was proving to be every bit as fruitful as Olduvai (and soon substantially more so), was a spectacular confirmation of the Leakey family's long-held belief, trumpeted loudest by Richard's aging father Louis, that *Homo* was an extremely old genus that went back as far as *Australopithecus* but, unlike the man-apes, was modern man's true ancestor. Skull 1470, which was found at Koobi Fora on the eastern side of the jade green lake and sits now in the vault room at the National Museums of Kenya (almost woodlike in appearance, with just somewhat pronounced eye ridges and forward-lurching upper teeth), was thought at first to be more than 2.6 million years old, or roughly a million years older than the *Homo habilis* at Olduvai. Such an age would have meant that *Homo* was older than australopithecines like the Taung child and close in age to the man-ape Lucy, which, at the low end of estimates, may have been only 2.8 million years old, the same age as *habilis.*

And this implied that perhaps *Australopithecus* was not really our true ancestor after all. The line leading directly to humans, Leakey believed at the time, might go back to 5 or even 6 million years ago. Because the history of *Homo* suddenly could be documented at least to nearly 3 million years, it meant, said Richard, that *habilis* was "living at the same time as some of our earliest australopithecines, making it unlikely that our direct ancestors are evolutionary descendants of the australopithecines—cousins, yes, but descendants, no."

That a specimen belonging not to the man-apes but to the genus in which modern humans themselves are classified could have existed back at the same time as the early man-apes (which Richard's father had often been quite alone in believing) was enough to throw paleoanthropology into a turmoil extraordinary even by its standard of chaos. Trees of evolution would once more have to be revised, this time as radically as anything in several decades, placing *Homo habilis* very near mankind's first hominid branching. The man-apes would remain peculiar and worth further study, but in the scheme of direct ancestry they might just become, as Louis Leakey had long predicted and fervently hoped, an evolutionary irrelevancy.

None of this would be much to the liking of *Australopithecus* proponents like Donald Johanson, whose Lucy (not yet unearthed) would have become one of those less relevant fossils.

Richard, who was twenty-eight years old when he made his remarkable discovery, proudly brought it back to Nairobi to show his father, whose elation was tangible: Here was the old man, seeing his fondest theory confirmed by a fossil discovered by none other than his own son! "Seeing and handling the '1470' skull was an emotional moment," Richard recalled. "It represented to him the final proof of the idea that he had held throughout his career about the great antiquity of quite advanced hominid forms." Louis warned young Richard to expect opposition, just as every one of his own major finds had provoked dispute ("They'll never believe you!"). He couldn't possibly have foreseen a fight, however, that was to become quite so personal and bitter.

Despite failing health, the older Leakey was about to embark on one of his regular journeys to lecture abroad, and the skull was shown him the morning before his departure for England. That was September 29, 1972. Louis Seymour Bazett Leakey, paleoanthropology's most legendary figure, died of a massive coronary two days later.

As he had warned his son, controversy quickly began to swirl around this new *habilis* discovery. At first it was a technical issue over the age of the tuff with which skull 1470 had been associated, but before too long it squared Richard off against White and Johanson.

The geological layer associated with the *habilis* 1470 is known as the "KBS tuff," named after one of Leakey's colleagues, Kay Behrensmeyer, a geologist from Yale. It was dated by what is known as the potassium-argon technique. In lava and volcanic ash is a radioactive substance called potassium-40, which breaks down into the gas argon at a constant rate—a geological as opposed to molecular "clock." By measuring the proportion of argon to potassium-40, one can give a layer of rock a fairly specific age. The first sample indicated an absolutely wild date of more than 200 million years due to contamination of the sample with older rock. But more samples were sent for analysis and more believable dates came back. In the case of skull 1470 and the KBS tuff, a refined version of potassium-argon dating showed the tuff, below which the skull was found, to be 2.61 million years old, meaning that the skull itself was at least that age and perhaps closer to 3 million.

The problem was that there were those who did not believe that date. It wasn't that Richard Leakey was exaggerating the age but rather that, like any such technique, potassium-argon is prone to giving different scientists different readings that can be interpreted various ways. Moreover, there is an entirely different way of dating geological tuffs, and that is by judging the age of a skull like 1470 based upon the fossils of extinct and previously dated animals with which it is found. In the case of the KBS tuff, the extinct animal that allowed paleontologists a yardstick of comparison was an ancient type of pig whose teeth were found below the KBS tuff.

That meant, of course, that these pig teeth were roughly the same age as the *habilis* skull. And one paleontologist, Basil Cooke from Canada, had determined from similar pig fossils found at Olduvai and the Omo region that the particular type of pig found at this level had lived "only" 2 million years ago. So how could the 2.6-million date for 1470 possibly be right? In addition (and in the end more consequentially), subsequent potassium-argon reanalyses indicated that the layer with which 1470 was associated was only 1.6 to 1.8 million years old. No big deal. That was about where *habilis* had previously been dated after the Olduvai discoveries, and meant the evolutionary tree would not have to undergo such a massive facelifting after all.

This all started to come to a boil in 1974, the year of the Lucy discovery. Picture this: two budding paleoanthropologists of about the same age, Richard Leakey and Donald Johanson, were both making extraordinary fossil findings at about the same time at sites in adjacent countries that each promised to exceed old Olduvai. Turkana was Leakey's kingdom, while Johanson was the crown prince of the Afar. Both fossil areas spilled forth all sorts of intensely interesting bones and both sites made their discoverers instantly famous. Their key specimens, however, in many ways negated each other. There wasn't much room for both Lucy and 1470—at least not at the extremes of age ranges they were given.

If 1470 had been 2.6 or 3 million years old, it would have been more important than Lucy in the sense that it was very nearly the same age (and at the most recent age given Lucy, older) yet was not a lowly man-ape like Lucy but of the *Homo* genus. And if it had been as old as originally tabulated, it would have shunted australopithecines like Lucy onto an evolutionary dead end. Who cared so much about Lucy if it was not really in the basal stock of human beings?

On the other hand, if (as turned out to be the case) 1470 was not as old as Leakey originally thought and Lucy was a million years its senior, then *Lucy* was much more dramatic in the larger scheme of things, in addition to the fact that she was a 40 percent complete skeleton as opposed to just a skull.

The new age established for 1470 was 2 million years or less, and in the view of many paleontologists it effectively knocked down the Leakey notion that an ancient version of *Homo* served as man's oldest and truest ancestors. At the same time it essentially reinstated the australopithecines, who, after all, had been around more than 3 million years ago. Johanson's man-apes from Ethiopia stole the spotlight.

The Leakey standpoint on ancient *Homo* could be preserved, however, if it could be shown that some of those australopithecines from Ethiopia were not really man-apes but rather were a type of *Homo* so ancient that they had been *misinterpreted* as australopithecines. Take them out of Johanson's category of *afarensis* and it was a new ball game. This is precisely what Richard and his mother (as seen in the previous chapter) had claimed: that *Homo* fragments were mixed in with the equally old australopithecines in Ethiopia. They claimed this not necessarily to preserve the Leakey ideology, as critics assert, but because there were indications that the australopithecines from the Afar Triangle indeed did have *Homo* traits—as Johanson himself had initially thought.

Johanson came, we already know, to the final conclusion that all his Afar samples were *Australopithecus afarensis,* and he couldn't have been too happy when Leakey publicly expressed the view that Lucy and her *afarensis* species were not ancestral to humans. Neither, in his turn, could Leakey have been endeared to rival Johanson's opinion that 1470 was younger than originally thought. There were even those who argued, and still argue, that 1470 wasn't really a *habilis* but an *Australopithecus!*

To make matters more volatile, Johanson had teamed up with Tim White, who wasn't getting along with the Leakeys. White had left the Leakey camp after accusing Richard of suppressing a paper that cast doubts on the original early date of 1470, only to have a second falling out with Mary Leakey during the subsequent footprint work at Laetoli. White didn't like the way Mary was running operations in Tanzania and in fact accused one member of her team of outright "incompetence." He also argued with Mary about the species designation *afarensis,* a debate she didn't much appreciate, not in her own workroom.

Add to this already potent brew the fact that Johanson had upstaged Mary at that Nobel Symposium and then had flagrantly attacked the Leakey family on personal terms. In a popular book entitled *Lucy: The Beginnings of Mankind* (1981), Johanson implied that Richard would have ignored the facts in any way necessary to keep the older date for 1470; that Mary had nearly fumbled the opportunity to study the footprints at Laetoli; and that the late Louis himself had been a stubborn, detested, eccentric fossil prospector whose early years were notable mainly for a knock on the head during a rugby match and whose later years were smudged by his womanizing.

Johanson also played up the fact that Louis and Mary had become lovers while Louis was still married, technically, to his first wife (Mary later felt compelled to discuss this adultery in her own book). Meanwhile, Richard was described as a delicate-featured man with a brittle personality, just as stubborn as his father and only informally educated at paleoanthropology (which implied that he made a poor spokesman for the field and that his claim to fame rested heavily on the fortuitousness of his birthplace and on who his parents were). The fact that Johanson had been funded by the autonomous L. S. B. Leakey Foundation (and had served in Richard's own foundation) had to have rubbed salt into the wounds, as did the fact that the Leakeys had given Johanson's new and closest ally in this battle, Berkeley's Tim White, his start as a major paleoanthropologist.

"Richard Leakey was made world-famous by 1470. His ideas about human evolution were solidified by it. But when its great age began to erode, he found that very hard to accept," wrote Johanson, slicing deeper still. "Instead of trying to reorganize his thinking in the face of growing evidence that the date was wrong, he defended the date. Finally he just drew back from it."

The fossil Lucy, said Johanson, slammed the door on the Leakey family's central *Homo* idea. Johanson also insinuated that Leakey may have undermined his efforts in Ethiopia, perhaps working behind the scenes to keep the American upstart and arch-rival out of Africa. When I asked Johanson why there was such tension between them, he replied: "I think in eastern Africa part of it has to do with the fact that there are a very limited number of sites and people want access to those sites and people derive notoriety and attention from working at those places. Until people like myself and Tim White began working in East Africa, at some of the more exciting places like the Afar and Laetoli, there were very few people besides the Leakeys who were working there. And I think there's a certain sense that we've invaded their territory. And there are some jealousies that develop from that sort of activity. There's a *tremendous* amount of controversy within the science itself."

Johanson has been known to describe paleoanthropologists as the "paleo-mafia."

White, who for the longest time could hardly contain his animosity toward the Leakeys, felt he had been punished for simply stating facts pertinent to 1470 too openly and honestly. He also felt the Leakeys took legitimate scientific criticism too personally. "They find it very difficult to accept the notion that one can criticize the dating of volcanic ash without criticizing the scientist who holds that dating to be the case," White told me.

Johanson's book incensed the Leakeys. In her own autobiography, *Disclosing the Past* (1984), Mary described Johanson's effort as a "lightweight book" fraught with questionable quotes, and she expressed nothing less than utter disdain for the way Johanson's field team had reacted to the discoveries in the Afar. In his book, Johanson had described his elation upon finding Lucy—the shouting, all sorts of hugging and jumping around— and had written that when more hominids were found the next year, the mood had been "a near frenzy."

When Mary read that, she found it "hard to picture a professional archeological expedition behaving in such a manner."

Her husband Louis had been in many fights, but this dispute, in Mary's view, was something different. "There were always disagreements. When my husband was alive he was always disagreeing with most people. He enjoyed it!" she recalls. "But I don't think they got down to personalities the way present people do. It's deteriorated, in that sense."

Her son Richard, who all but ignored the Lucy allegations in *his* autobiography, *One Life* (1983), despite the damage to his reputation, expresses extraordinary bitterness over the entire matter. Though not always one to avoid contention, he is bone tired of everything from the *afarensis* controversy to the one over his KBS tuff. It even made him think of quitting fossil hunting.

"I talked about getting out to a certain extent because I became very, very depressed and unhappy about the personalization of differences of opinion about fossils," says Richard, a casual yet intense and sharply featured man who, besides contending at such an early age with the KBS uproar, also had to undergo the dangers of a kidney transplant. "And I found the conflict over *afarensis*—the way that became—sordid. I didn't see why I was being attacked personally and my integrity was being openly discussed at a level that was irrelevant to the science. I just didn't like [the book *Lucy*], and it has continued. There's a great deal of antipathy being addressed towards me for reasons that I honestly don't understand because I don't share them. I was very happy to have helped [Tim White] get his career started. We were good buddies. Just because we think the fossils are different, I don't see why we have to hate each other. And I don't want to hate him. That's why I want to get out of it."

Although it came with the turf, there was ultimately something embarrassing about seeing the titans battling it out. Both Johanson and Leakey were exceptional men. Richard, though never extending his formal education, was as bright a paleoanthropologist as any university could hope to produce, knowledgeable about the intricate aspects of just about any stage of hominid evolution, a wizard at field organization, the most widely cited fossil hunter of the century, his name all over the journals, outdoing his own legendary father; and Johanson and White were masters at morphology, not only dispensing extraordinarily useful information and new finds (like their discovery of another *habilis*-like humanoid in Olduvai during 1986) but also offering those with whom they came into contact relevant criticism, an equally comprehensive knowledge of the entire field, and an infectious enthusiasm.

And then there was the pathos of seeing Mary having to fight it out as a woman in her seventies. This was a clan that had made it through more than half a century of fossil hunting by unearthing one find more extraordinary than the previous, weathering, along the way, attacks by killer ants, close encounters with rhinos and crocodiles, a lion that one night smashed the glass in Mary's hut and filled the window with its head, destruction of camp facilities by marauding natives, bouts of malaria, puff adders at Laetoli, steep cliffs from which a Land-Rover could fall, and "roads" that were really nothing more than rivers of the thickest imaginable mud through terrain where the lost may never be located—surviving all this only to find, at the end of the road, their credibility pockmarked by salvos fired with megaton force from the new citadel of evolutionary studies, the distant city of Berkeley.

Mary has retired, but Richard loves the desert and will in all likelihood remain in the field, at least part-time, despite the ruckus. "I enjoy life, I get a tremendous joy out of being alive and being human," he says in a clipped and precise British accent. "And I don't really care, in the final analysis, whose bones are older than whose."

Well, maybe, but what Richard did after the decade-long dispute that demoted his 1470 was head back to western Kenya and find an ancient skeleton that was even more complete than Johanson's mind-boggling Lucy and represented the next major stage in human ancestry after *habilis*, a specimen no one who believed in evolution could conceivably deny was an early human: *Homo erectus*, who brought us a big step closer to Eve and who was the greatest ape-man of them all.

12

IF PEKING MAN was destined to remain the most famous *erectus* and indeed one of the species' very type specimens, the most complete and among the oldest *erectus* was to come not from China but from East Africa. It was found in 1984 by a Leakey employee named Kamoya Kimeu, who spotted a fragment of frontal skull poking from the moonscape near Lake Turkana.

The discovery was significant not only because it emphasized that Africa may have been the origin of *erectus* just as it had seemingly been the origin for man-apes and *habilis,* but also because *Homo erectus* was really the first humanoid that modern man could identify with. Almost no one disagreed that *erectus* belonged to the genus *Homo.* It was *erectus* who later fathered both Neandertals and a line of archaic humans who led to Eve, wherever *she* originated.

Kimeu had made the discovery next to a small river called the Nariokotome. Not far away, Turkana's jade waters, now in retreat, glistened across a huge expanse in the Great Rift Valley, hiding more secrets and cooling the air that shimmers up from the volcanic debris. That is where the skull fragments of the *erectus* skeleton were: in yellowish-brown siltstones and sandstones, hundreds of them, soon to form most of the calvaria, or upper part of a skull, more and more pieces discovered as Leakey's exploration team painstakingly sifted the earth, which yielded a mandible, rib bones, vertebrae, femurs, several isolated teeth, and the

Homo erectus

clenched, vacant, and wide-eyed skeleton of a face—pieces of an incredible jigsaw.

Richard Leakey was in his forties now, alert and full of energy after the KBS uproar and the frightening kidney transplant. He was as immersed in fossils as his parents once were, despite those earlier years of edging away from them. By now Richard's name was a frequent by-line in *Nature*, and his reputation was expanding to the point where this self-taught man who might have felt out of place at an elite and confining school like Berkeley, this man of English descent who was nonetheless a proud native Kenyan, this man whose boyhood had been spent at Louis's camps, bathing in remote crocodile waters, was becoming one of the more famous scientists in the world—if not *the* most famous. His critics could criticize his ego and look down their snoots at his lack of an academic background, but in reality Leakey was less pretentious than many lesser-known scientists, and no one could look down at the bones he kept turning up.

They especially could not dismiss this find at Nariokotome. It was unheard of even by Leakey standards. Nearly a complete fossil skeleton! No one had ever found more bones belonging to a single pre-*sapiens* individual. And if it was less dramatic than Lucy because it was younger, it was in other ways more noteworthy, especially in that many paleoanthropologists believed it was of the type that served as Eve's immediate ancestors. Where most fossils are composed of only a lower jaw with one or two intact teeth, this skull and mandible, in the end, were missing only tiny fragments, and the rest of the skeleton was minus only hands, some neck vertebrae, and feet. Every tooth was present. The ribcage and vertebral column were almost complete. By comparison, Lucy, who had just bits of the skull (though spectacular pelvis and ribs), looked as if she'd been run over by a freight car.

It was a boy, this new fossil, a boy who had died around 1.6 million B.C. And he seemed to be part of a population that had preceded the *erectus* in Asia. Other fragments found near Turkana registered at about 1.9 million, which would make them twice the age of the famous Java *erectus* and perhaps three times as old as Peking Man. Once more a Leakey was transforming prehistorical knowledge. And the picture was the enormously exotic one of beetle-browed, club-wielding first True Man vying on the savannah with the big-jawed and flat-faced man-apes that, unlike *erectus*, who was so clearly *Homo*, looked like they belonged to another solar system.

Indeed, the Nariokotome Boy, as he was soon called, was a quantum leap up from the alien-looking australopithecines. There he was to see, scattered in a curving trail of bones well into a hillside near Turkana's western shore, with nearly twice a man-ape's brain volume and standing at nearly five and a half feet. He had been only twelve or so when he died; they could tell that by his bone "epiphyses," which had not yet fused. Had he lived a few more years, he may well have ended up a strapping six-footer—the first ape-man who could have played basketball.

"Incredibly, he was washed away in a flash flood, drowned or something," says Meave Leakey, demure and attractive, wife of Richard, showing a visitor the cast of this skeleton in an anteroom of the Nairobi museum. "And then the river deposited him in a marshy area. His bones were not disturbed by scavengers or anything. It's what every anthropologist *dreams* of finding. First Kamoya found just a few skull fragments, and then we started sieving and then found more skull fragments, and then after that we started excavating, and once we got to the main area where the bones were, they were coming out *so* often. We were finding so many bones. The initial year we excavated about six weeks and the following year three months and the following year three months. Most were found in the first two years. It was really exciting. At night you'd dream about bones. You were always dreaming about bones. We were the first people to see this person become *erectus!*"

Here we were, finally, to Eve's own great-grandparents, the line that led, more than a million years later, to the kind of female who'd served, according to Cann, et al., as our own grandmother in 200,000 B.C. Somehow out of the nearly indecipherable menagerie of hominids and humanoids had arisen a mortal who may have been awfully uncouth in style and looks but was clearly more human than *habilis*, with a less protruding face but prominent browridges and a sloping forehead. If there wasn't much in the way of a chin, the brain was suddenly immense—775 to 1,225 cubic centimeters—and overlapping the range of 1,000 to 2,000 cubic centimeters in modern humans.

Erectus was indeed an intriguing character, and though it is always dangerous to speculate on lifestyle, with these ape-men the urge is irresistible. They may have had rituals. They probably built crude shelters and slept on hides or leaves. They certainly had fire, and lots of it, which meant they may also have cooked. At Swartkrans in South Africa is evidence of controlled flames 1.5 million years ago. There are also heaps of ash in Chinese caves. Perhaps they first noticed fire from that lightning striking a forest, from a volcanic eruption—or from chipping flint tools with iron pyrite and causing sparks to fly into those bedding leaves. Once embers were going, they may have banked them in sod to preserve them. Besides warmth and cooking purposes, the fire may have been used by *erectus* to harden the ends of digging implements and perhaps even the sharp ends of the first protospears.

If so, that was only part of their tool kit, which included choppers, chisels, awls, hammerstones, anvils, scrapers, and flat, pear-shaped quartzite tools with long, sharp cutting edges. These are known as the hand axes. They were to spread the world over, lingering right through a transitory stage of archaic *Homo sapiens* and into the advent of modern men. For all anyone knows, an increased use of scrapers may indicate the fashioning of hides and leather—the first real clothing.

Erectus fought and hunted and did both with what can only be described as savagery. At a hand ax (or "Acheulian") site in Olorgesailie, Kenya, are indications that they killed giant gelada baboons by smashing them over the head, then butchered them by twisting the thigh bones, striking the arms without mercy, and amputating the shoulder joints by crushing the proximal humerus—tearing apart the animals from limb to limb in what may have been a seasonal or ceremonial event—a *rite de passage* of slaughter.

There were hints of other rituals and perhaps spiritual rumblings. "Self-awareness and the ability to reflect on the universe may well be tied to the expansion of the brain seen in the fossil human ancestor *Homo erectus*," wrote Leakey. "Though we can never prove it, if this were so, 'religion' could be a million years old."

So could monogamy. There wasn't the extreme dimorphism in *erectus* that there was in certain man-apes. But then, humans weren't gibbons. Every day may have been a sexual free-for-all. That seemed appropriately chaotic and bestial, and ape-men were certainly bestial. Still, for all anyone knew, *erectus,* over the course of its million-year existence, developed the first rituals pertinent to pair bonding. No one was suggesting they said vows, but morphological indications from places like Java showed the capability (see Notes) of limited speech. It wasn't as if they could really talk, but there did seem to be some communication going on: from the high plains of Gadeb in Ethiopia are indications that *erectus* may have carried material for obsidian hand axes from a source sixty miles away, which would take some intellectual coordination, and certain scholars have guesstimated (as just about everything above is guesstimated) that a clan of *erectus* may have ranged over a territory of 300 miles for its meat. In a day's drive you might have spotted two or three small groups of them, cowering from your Land-Rover.

The ability to make fire gave them greater range still, allowing them to expand into the cooler north—in the direction of the periodic, encroaching glaciers. In China one might have been lucky enough to catch them doing battle with *Gigantopithecus,* the monstrous ape (see Notes), or with a cave bear. There were also rhino to watch out for, even in Asia, and perhaps, too, hostile *erectus* clans. Whether in ritual or as simple savagery no one can tell, but *erectus* seems to have been a cannibal, smashing or otherwise enlarging the base of skulls to get at the tasty (or power-giving) brains at places such as Zhoukoudian.

In 1989, Milford Wolpoff, one of the world's foremost experts on *erectus* and professor at the University of Michigan, proposed that the increased thickness of *erectus* skulls may itself have had something to do with violent behavior. "I view it, along with the expanded cranial buttressing system that appears particularly well expressed in the Asian populations, as related (and perhaps responding to) the high level of healed cranial injuries that seem almost ubiquitous in *Homo erectus* crania," wrote Wolpoff for a publication in France. "The causes of these injuries remain speculative, perhaps hunting accidents in reflecting the inability to kill at a distance, but more probably these are the result of elevated magnitudes of interpersonal violence especially given the predominance of injuries on female crania."

Their longevity could only be guessed at, probably at the very most forty to fifty years. And so once again the words of Becky Cann came echoing from the lab: nasty, brutal and short was the way she had characterized the lives of Eve's forefathers. And if so, there were also the more delicate moments. For all anyone knew, *erectus* may have established the precursors of art. There's no real proof of that, and it's probably not even true, but there are those who cite what they see as hints of pigment use in the Gadeb. Although it seems mighty unlikely that there were any cave paintings as yet, or gourmet gazelle, or actual matrimony, *erectus* certainly knew how to do one thing, and that was to survive: from sometime before the Nariokotome Boy to about 250,000 years ago there were still pockets of them in places like Asia—a span of more than a million years.

Though *erectus* was still somewhat apish of face, the nose was rising, however, out of the flatness of the man-ape's, there was most likely no more fur (though plenty of hairy backs), and *erectus* would probably have had the wits and coordination to drive a car or indeed play a sport—given his rugged physique, more likely football than basketball. He was muscular as all get-out. A female *erectus* would have been a professional wrestler. And whatever the sport, there might not have been much need for a helmet, given the thickness of the skull. It was a mass of bone that looked, from the top, like the shell of a turtle. There were the hulking browridges and a face that still thrust forward enough to ring vague memories of a snout.

This is the species that Carleton Coon said is resembled by living Australian Aborigines, who are still in the process, he claimed, to the horror of his colleagues, of shrugging *erectus* traits off. Statured the way it was, *erectus* may have been able to walk a busy street without creating a scene (if properly attired) or into a subway car without causing a mass exodus. He was certainly much less of an oddity than an alleged human like *habilis*. Leakey saw this

firsthand while helping the British Broadcasting Corporation produce a series on man's evolution. "When we were making the film,
we took some face masks and we modeled them onto actors' faces
with silicon rubber and you could not tell that they were masks,"
recalls Richard. "But anatomically they were correct for erectus in
that we lost the forehead, we built up the browridges, and we gave
them a prognathic [projecting face] profile. These masks took
several hours to put on each morning, and whilst waiting for the
sun to appear, these actors would be standing around in clothes
smoking and talking, and crowds would come by but most onlookers were unaware. It was only when someone suddenly started
looking that it was appreciated. In a subway, chances are they
wouldn't be noticed."

But if they were, they certainly would have drawn that
second glance, because erectus was still pretty darn primitive, and
some artists even portray this ancestor with the head of an ape and
the body of a human—which hardly brings to mind an Aborigine.
That's how its very discoverer, the Dutch anatomist Eugène
Dubois, pictured erectus. Dubois, upon finding Java Man, we
should recall, thought the skull represented the perfect intermediate between man and ape, resembling the cranium of a huge
gibbon.

Never before had there been seen so flat and low a human
skull. There was no question major differences existed between
this ancient population, which Dubois had called Pithecanthropus
erectus, and what we are familiar with anywhere today. The lower
part of the squama temporalis retreated outwardly, just as it does
in apes, and one of the teeth looked like an orangutan's. At the
end of his life, Dubois was dragged away screaming, in fact, that
erectus was more ape than he was human.

But Dubois didn't have the benefit of australopithecine
fossils, which make erectus look so very human when the two
species are compared, and he had no idea that a two-footed gait
went back millions of years before his Java fossils, for if he had,
surely he wouldn't have given it the inappropriate name erectus,
implying that it was the first to walk erect.

Of another erectus discovery in 1975, which was made by
another member of the highly skilled Leakey team, Bernard

Ngeneo, Richard wrote, "What a find it was! There was no doubt that this was not *Australopithecus* nor even *Homo habilis* but rather *H. erectus*, a more immediate ancestor of ourselves. Words are inadequate to describe our feelings because for the first months we had suspected that *H. erectus* had lived in Africa more than a million years ago and here at my fingertips was proof, a perfectly preserved skull found *in situ* in sediments over 1.5 million years old."

The 1975 *erectus* and the subsequent discovery of the even more spectacular Nariokotome Boy were the culmination of many decades of *erectus* discoveries in Africa. Richard's parents also had made important such finds. In 1960, Louis had come upon an *erectus* or *erectus*-like fossil at Olduvai that became known as "Chellean Man." It is now thought to be about 1.2 million years old, which implies that it too was around when australopithecines were still alive, since the robust man-apes are thought to have existed until about a million years ago. The formal name was *Homo leakeyi*. Even further back, in 1935, Mary had found scraps of fossil that are now considered to have been *erectus* as well. When one considers the *erectus* specimens that spanned from places like Swartkrans in South Africa to Ternifine (now known as Tighenif) in Algeria, the image is of a continent fairly teeming with ape-men.

If the old ages assigned to *erectus* fossils in Africa indicate they had originated there, this is not to say there haven't been hints—if highly questionable hints—that at the same time as the Nariokotome Boy, or even earlier, *erectus* was also elsewhere. There were those who'd believed that some upper right and left central incisors from Yuanmou County, Yunnan Province, in China were 1.7 million years in age, and others who thought certain Java specimens went back nearly that far too. Some said there had been pre-*sapiens* activity 1.75 million years ago at a location in southern China known as Shangnabang. Still others believed *erectus* had been in Europe close to or even more than a million years ago in such places as Spain, Hungary, Italy, and Germany. According to British paleoanthropologist Bernard G. Campbell, author of the seminal textbook *Humankind Emerging*, a cranial fragment near Orce, Spain, was given a highly tentative 1.3 million date. Campbell also reports in the 1988 edition of his book that tools 1.8 million years old—as ancient as what was found at Olduvai Gorge—have been disinterred from Chilhac, France.

But the best that can be said of all these dates is that they are tentative in the extreme. In the case of Yuanmou, for example, the 1.7 million figure seems to have been based on faulty interpretation of the geomagnetic polarity and a misreading of the biostratigraphic record. Yuanmou may be a third that age. The earliest reliable fossil remains from China seem to be from at least 900,000 years ago in Gongwangling. At their youngest, certain of the Zhoukoudian or Peking Man fossils might be only 230,000 years old.

As for Java, the oldest there are perhaps 1.3 million. After Dubois's disclosure of an *erectus* skullcap near Trinil in central Java a century ago, more skull fragments—many more—would be found in a region to the west known as Sangiran. All told, Java would yield dozens of intriguing skulls, including some that defied easy classification. But "the earliest hominids (except possibly *Meganthropus*) in Asia all belong to *Homo erectus* which is now dated at 1.3 million and quite possibly much younger in Asia," writes Geoffrey Pope of the University of Illinois, who makes regular treks to China.

Around that same time, according to others, there are nebulous and certainly less than reliable indications that *erectus* may also have been in Europe. At Isernia in central Italy, assemblages of crude tools (minus hand axes or any bifaces) have been dated at perhaps 750,000 years old by radio-potassium. The same is true at Karlich, Germany, where paleomagnetic dating, in which rocks are dated by studying the lineup of magnetic crystals (which indicate known periods during which the earth's poles have reversed charge), shows more than 730,000 years ago.

Many paleoanthropologists do not accept the existence of any classic *erectus* in Europe, and so what we are left with is the likelihood, based upon the best current evidence, that *erectus* originated in Africa and then, around a million years ago, wandered through the Middle East (where evidence of *erectus* has also been unearthed) all the way to China and Indonesia—there to become known as the Java and Peking men. These were the beings who, most scientists agree, were the first ancestors to establish themselves out of Africa, "setting up seasonal camps in southern Asia," added Rebecca Cann, "and later in northern Eurasia."

Once asked why *erectus* would leave Africa for such far-flung parts after so many years in Africa, Milford Wolpoff said: "They left when they did because they wanted to, because they had to, and especially because they could."

The expansion from Africa to the islands of Indonesia meant a movement of about 7,000 miles. This expansion seems to have come at a time when Indonesia was part of the Asian land-mass. *Erectus* seemed to like warmth, staying away, for the most part, from the glaciers that were periodically covering much of northern Europe during this the beginning of the Ice Age.

"Considering climatic variation in the last 600,000 years, it is now well-accepted that the ebb and flow of glaciers encroach-ing on Europe and Asia has periodically driven our ancestors to take refuge in Africa, the Middle East, and perhaps India," said Cann.

That granted us the qualified view, for now, of *erectus* origi-nating in Africa sometime before 1.6 million years ago and spread-ing up through the northern part of that continent and across the Middle East to some of the farther reaches of Asia before being replaced much later by a second wave of more advanced humans—Eve's children—who, according to Cann, et al., rose from that *erectus*-like band in the sub-Sahara and took over Asia and the rest of the Old World from their primitive, unevolved relatives.

Getting back to the original migration of *erectus*, it was Bernard Campbell's view that the expansion into new habitats was accomplished by what he calls "budding." A few individuals would split from a group and form their own autonomous band on nearby land, and then their children and grandchildren would do the same. It wasn't as if whole troops marched great distances *en masse*. Instead, it was like the spread of crabgrass.

Here this generation, there the next. At the time, as I said, Java was possibly connected to the Asian mainland. The oceans were lower because this was the period of global coolings and water got locked up in polar ice caps. Africa for all we know was linked to Europe across Sicily, says Campbell, and to Asia through Egypt and Ethiopia. "*Homo erectus* probably drifted back and forth throughout the tropics, with new bands branching off from more settled populations as their numbers increased," he wrote. "Some dispersed north from Java into China; others spread north from Africa across land bridges to Europe, or entered Europe after skirting the Mediterranean by way of the Middle East, Turkey, and the Danube."

Archeologist Lewis Binford of the University of New Mexico expresses the spread of *erectus* somewhat differently, saying that tools similar to the technology used in Africa as far back as 700,000 years ago appear in China by about 450,000 B.C. and that Russian artifacts indicate the cultural spread did not go through India, "but across the Turkish plateau or out of the Black Sea basin along the west side of the Asiatic mountain range and into China through the Gobi Desert or some big open place." Binford says there appears to have been a "major radiation" of *erectus* into southern Asia around 300,000 to 375,000 years ago, one that also included Europe.

As simple and clean as such an interpretation may seem, it could spark any number of intense debates given the presence of more than one paleoanthropologist in a single room. As was the case with every other species, there are questions not only of dates but also of basic morphology: what one scientist proclaims as an *erectus* may look like something else to another morphologist, especially if that morphologist is a competitor.

There was even a question of whether, instead of being called "*erectus*," these beings might more accurately be considered a type of *Homo sapiens*—a "grade" or loose collection of populations all evolving toward more modern humans. At the other extreme were a few *erectus* fragments that looked more like *habilis*.

"I hope we would all agree on one thing, and that is that there is not much consensus around this table," David Pilbeam once said during a conference at Karlskoga, Sweden.

"I think we have reached a point," added Glynn Isaac, "where it is almost antisocial in our science to present only one hypothesis."

Only one thing was for sure: there was a pretty good collection of *erectus* bones, especially from the caves at Zhoukoudian (formerly Choukoutien). There were so many bones, in fact, that drugstores and street merchants in nearby cities used to grind them up and sell them as potions and aphrodisiacs. The name of the place where many of the fossils had been located was Dragon Bone Hill, and as early as 1937, the Peking Man site had yielded several dozen individuals depicted by five complete skulls, nine fragmentary ones, six facial fragments, more than a dozen lower jaws, and 152 teeth. Additional specimens from modern eras would be found in what is referred to as the Upper Cave. It was here that Franz Weidenreich did his monumental work detailing the characteristics of *erectus* fossils (which were originally called *Sinanthropus pekinensis*), and it was also here that those signs of fire were found—ash, charred wood, and burnt bone heaped twenty feet high.

Meanwhile, from Lontandong Cave in Hexian County has come an *erectus*-like specimen with the wholly infatuating hint (in terms of higher forehead and thinner cranial bones) that *erectus* in Asia was turning modern.

The great barrier posed by the Chinese language and the large cacophony of specimen names muddled the issue in Asia and also in other *erectus* locales. Harvard's Ernst Mayr, who is in his eighties now and has been described as the world's greatest living evolutionary biologist, surveyed the disarray of various names and helped spearhead an effort to bring these and other seemingly artificial categories for *erectus* into one taxon. Aside from the use of genus designations like *Pithecanthropus* and *Sinanthropus* for the Asian finds, there were those who insisted upon arcane species names such as *heidelbergensis* and *modjokertensis,* or in South Africa, *Telanthropus capensis.* "I pointed out that it really wasn't sufficiently different to be called a different genus," Mayr explains. "So instead of *Pithecanthropus erectus,* I called it *Homo erectus,* and in due time it's been accepted by just about everybody."

It was Wolpoff's contention that physiological characteristics evolved gradually over *erectus*'s existence, the posterior tooth size continuing to decrease while the posterior braincase increased. The average cranial capacity for later *erectus* may have been 30 percent greater than for earlier assemblages, Wolpoff claimed. There had also been an increase in orbit size, says Wolpoff, which could reflect an increased number of the eye's rods and cones and thus higher visual resolution.

But there was hot debate about this concept of gradualism. Others thought evolution stood still for long periods of time and that change occurred in dramatic fits and starts—a concept known as "punctuated equilibrium," promoted most recently by biologist Stephen Jay Gould. Could anyone agree on anything? If it wasn't a battle over names, evolutionary trees, measurements, anatomical observations, tooth-wear patterns, lifestyles, or prehistoric tools, it was a battle over general philosophy and always, always a battle over dates.

But few quarrel with the general idea that in one way or another *erectus,* after going through a period of transition, led to Eve. How could it have been otherwise? At Turkana, in addition to the Nariokotome Boy, was another fossil impression—a fig leaf.

13

Eve's entrance may have been somewhere in a hodgepodge of primitive people who, depending on which paleoanthropologist you consulted, were either advanced forms of *erectus* or belonged in a new, ambiguous, and catchall category of "archaic" or incipient *sapiens.*

Successors to Peking Man and the Nariokotome Boy, archaics had all the earmarks of a transitional type, and though time ranges for their existence are stated only at the peril of those who attempt such definitions, they probably existed, these half-*erectus,* half-*sapiens,* on the order of 500,000 to 100,000 years ago.

That meant they had appeared on the scene before the last of the conventional *erectus* disappeared, and like them, the archaics lived in both Asia and Africa. Not far at all from where the Peking and Java ape-men had been found, pockets of humankind seemed to be emerging from the *erectus* morphology, subtly shedding, like the specimen from Hexian, those features or nuances that tied them strictly to the old ape-man.

On the banks of the very waterway where Dubois had collected the Java *erectus* were also found skulls, jaws, and teeth of what Franz Weidenreich believed were just such links between *erectus* and modern men. The Java fossils, discovered between 1931 and 1932 by natives working for the Dutch Geological Survey and thought possibly to have resulted from a long-ago cannibalistic feast, appeared to be descendants of a population resem-

An archaic

bling the *erectus* of Sangiran. Dubois himself thought the skulls, once known as *Homo soloensis* but now more noncommittally known as Ngandong Man, were ancestors of the Aborigines. He described them as "proto-Australian."

If that was true it meant that the fossil record from Asia was developing a direct continuum from humans who had lived a million years in the past right up to transitional archaics and modern humans. In a limestone cave which Chinese farmers had stumbled upon while digging for phosphate fertilizer in the Kwang-tung Province north of Canton, there was a skullcap that was interpreted by some observers as just such an intermediate speci-men, and in 1978 near Dali in a valley of the Luo River in Shensi was yet another specimen which presented, with its low vault and heaped browridges, what Harvard's W. W. Howells described as "a mosaic of archaic and progressive features."

Everyone seemed to be stumbling over remnants that, with their progressive features, didn't quite fit the *erectus* classification. It was as if the archaics had been much more numerous in popula-tion than their forefathers, or had grown more adept at adjusting to widespread terrains. They existed in the desolate mountains of Russia, in India, in the Middle East, and, as always, in the eastern and southern parts of Africa. An especially old braincase was found near Saldanha Bay in South Africa, and a mandible from Baringo, Kenya, resembled a specimen from Sidi Abderrahman in Morocco, showing important divergences from *Homo erectus*—showing, too, that the entire length of Africa had been traveled by *erectus*'s progeny. Another Moroccan fossil near the village of Salé gave indications it was 400,000 years in age. Indeed, Morocco and Algeria, both in northwest Africa, seemed to be stopping off points on the way to Europe (if indeed Africa was the origin) or at least longtime settlements once man-apes had attained their manhood, for *erectus*-type fossils were known from a good number of locales up there, the most famous of which were some lower jaws from that place in Algeria named Ternifine.

There were a minimum of a dozen such examples from Africa, including a broken lower jaw from a site known as Cave of Hearths in southern Africa (found with hand axes) and the famous 1921 discovery of "Rhodesian Man." The latter was from a cave in a knoll known as Broken Hill in Zambia, and it bore resemblances both to the archaics from the Far East and to a specimen found at Lake Ndutu near Olduvai.

It also bore resemblances to a good number of archaic specimens found in Europe, where perhaps thirty examples of this vague intermediate type had been found from Mandrascava, Sicily, and Lake Banolas, Spain, to Ehringsdorf, Germany, and the River Thames in England. And such a tally was without counting the Neandertals, who came later and were also early, brutish forms of Homo sapiens.

It is worth running down more of these sites to develop the sense that except for Australia and the Western Hemisphere, the archaics were pretty much all over the place. Of about the same age as the Ngandong fossils in Asia, for example, were remains of huskily built humans found near Heidelberg, West Germany, in Petralona, Greece, at Bilzingsleben, East Germany, in Vértesszöl-lös, Hungary, and at the eastern end of the French Pyrenees. There, a place called La Caune de l'Arago yielded a facial skeleton and isolated teeth that were associated with the same type tools that had been so exhaustively used by the *erectus* forerunners. The French and Greek specimens shared striking similarities with Rhodesian Man and were also compared to fossils found at Mela Kunture and Bodo in Ethiopia. Among other new characteristics, it looked like man had started to form a chin.

Surely our Eve must have had a chin, and the chronology for archaics automatically made them candidates for the mito-chondrial theory of origin. The estimated dates for archaics ranged wildly—Petralona has been variously estimated at anywhere be-tween 200,000 and 730,000 years—but in general this first wave of archaics, sharing many features with one another, seems to have existed 500,000 to 100,000 years ago, as did the DNA Eve.

As far north as England humankind was lightening up on the *erectus* traits, gentler in the curves of skull, bearing remarkable resemblances, in certain instances, to primitive *sapiens*. Could Eve truly have been in there somewhere? Was this one mystery that the fossil record, alas, was going to solve?

What surprises were in store! For now suffice it to recall that Cann, et al., had said, based on the constant 2 to 4 percent rate of DNA divergence, that Eve existed in 283,000 to 140,500 B.C., rounding it off at 200,000. Such a range of dates meant that this prodigious mitochondrial matron lived, at one extreme, during the tail end of *erectus*'s tenure, or, at the most recent end of the range, at very close to or perhaps just after the point at which the fossil record showed archaics turning into the modern *Homo sapiens sapiens.*

If she came during or after this transition, then she was indeed a member of the first anatomically modern *Homo sapiens sapiens,* who, according to the DNA data, lived somewhere in Africa between 200,000 to 100,000 years ago. If not, she existed as one of the last of the archaics.

We'll assume for now that she was an archaic—that's what Allan Wilson was calling her. "One consequence of the pattern of inheritance of mitochondrial DNA is that the mitochondrial Eve will almost always have existed a considerable time before a newly derived species becomes established," he had said. "For this reason, we see it as probable that the 200,000-year-old African female from whom we believe we all derive our mitochondrial DNA was a member of the archaic *sapiens* species, and was not yet an anatomically modern human."

If you believe that was when our common mitochondrial ancestor lived, forget the sweet images of Adam's "fairest," junk the Dürer painting, and ignore any dainty little fig leaves. She would have been a brute. Her brows would still have had the unsightly thick ridge. Her head would have been rather low and sloped, with not much of a forehead to caress—and no chin. She would have smelled like a goat. And if she got mad, she may have whacked Adam over the head with a hand ax.

That's what was meant by "archaic." A semi-apeman. And it was quite at variance with the catchy romantic image Wilson and his colleagues had provoked when they won headlines by referring to their "Eve." If what Wilson was saying was true, she may have been inferior in an aesthetic sense even to the lowly Neandertals, who were not on the scene yet but also rose from this same "evolutionary" mix. In fact, certain scholars believed that some incipient Neandertal features were already becoming evident in the French and Chinese specimens as well as those from Ngandong. And whether or not the simple sight of her would have caused a mass exodus from a subway (if she was dressed, as Leakey said, she may not have been immediately noticed), a close look and a close whiff may have been enough to cause even the tottering wino to find another seat.

The disappointment at her being an archaic, if archaic she in fact had been, was comparable to discovering there wasn't a tooth fairy.

Wilson of course never had said there was. He seemed simply to be trying to tie in his results with the fossil record, instead of quite so directly bucking the paleoanthropologists as he had during the 1960s when his and Sarich's gorilla-chimp-human results with albumin had been ignored or otherwise received negatively. No matter. There was going to be a gale of contention anyhow. But placing Eve as an archaic, between modern humans and *erectus,* was, for now, a reasonable compromise. It was also hedging his bets and providing something of an escape route, even if it did do major damage to the Eve image.

Even with all the fossils found throughout the world, there was not a consensus on what an archaic was. Paleoanthropologists had a wrenchingly hard time deciding if such remains should be placed into this all-purpose taxon of transitional archaic forms or whether they instead should still be viewed as *Homo erectus.* Whether humans of this period were *erectus* or archaic *Homo sapiens* depended, again, upon which paper you read or which textbook you bought. And the interpretation, in this highly subjective field of science, was as often as not in accordance to a scientist's own previous discoveries or equally cherished theories.

A case in point was the *heidelbergensis* species mentioned briefly above and in the preceding chapter: the specimen from West Germany. For many paleoanthropologists, this sample—consisting of a lower jaw and teeth discovered at a place called Mauer—had to be an *erectus* because of its date, which some believed went back to 450,000 years ago. It had no chin, as *erectus* had no chin, and its jaw was positively massive.

But in the massive jaw was a startling contradiction: small, modern teeth that hinted at a divergence toward *Homo sapiens.* Many paleoanthropologists continue their insistence that there is as yet no definitive proof that *erectus* lived in Europe, and so the Mauer specimen, despite its primitive state, is sometimes lumped into the limbo of the "archaic" category, along with a plethora of specimens from such places as Montmaurin, France, and Atapuerca, Spain, whose status is still debated and which likewise find themselves at this boundary—this no-man's-land, to skew metaphors—between *erectus* and *Homo sapiens.*

In India was a specimen that Milford Wolpoff said looked Chinese. He categorized it as "late *erectus.*" Others said it was "early modern." For a while it looked as though teeth collected by the Leakey colleague Kamoya Kimeu (along with Richard's brother Phillip) at Lainyamok, northwest of Lake Magadi in Kenya, might represent the oldest-ever archaic (at 700,000 years old); but reevaluation has led one of those who analyzed the site, Pat Shipman of Johns Hopkins, to conclude that it is probably just an *erectus.* "The closer you look," warns Wolpoff, "the messier it gets." And it was very messy here indeed. The problem for the Berkeley geneticists was that there were no genes to put into an agarose gel. Scientists at the University of California at Los Angeles had developed a technique to analyze bone collagen carbon and nitrogen isotope ratios in prehistoric men by use of electron microscopy, but that had been only to find out what certain of these people ate—their diets—and so far such analysis went back only to the Neolithic, or "Later Stone Age," which meant only a few thousand years ago. With archaics we needed to go back to what paleontologists refer to as the Middle Pleistocene and what archeologists, to make sure this too is no simple matter, call the "Old Stone Age" or Lower Paleolithic.

Since it was during a time of glaciation, which peaked about every 100,000 years, it is also known as the Ice Age. The scenic backdrop for entrance of the archaics was essentially a world of temperatures colder than the average weather today during the glacial periods and yet substantially warmer—with water buffalo thriving as far north as Germany and monkeys chittering in Italy—during long respites sandwiched between those periods of bracing cold. If Africa remained more hospitable than regions north (though with reduced temperatures and thus more rain), in Europe the animals—wooly rhino and wooly mammoth—said it all. There were also musk oxen, reindeer, and the arrival of the modern horse. And in summer, as the wind picked up debris collected by glaciers, the whole of a horizon might be yellow from an advancing dust cloud.

In fact, the period 500,000 to 100,000 years ago—about a million years after *erectus* had first appeared in Africa and again around the time the last full-fledged *erectus* disappeared from China or from wherever it disappeared—was about as cold as the Pleistocene got. There was one ice expansion after another. And during the severest ones, ice spread from the Arctic over much of North America and Europe, covering northern England, Germany, Scandinavia, extending down in some cases nearly to Moscow, and causing patches of white in highland France, the mountains holding their snow through summertime, temperatures in July at times dropping to freezing, and central Europe turned into tundra before the cold retreated with the onset of a warmer period known as an interglacial. Even in Spain the ground bore traces of frost patterning like what you'd find in Alaska.

But the humans from this time period had fire, they had hides, they had rudimentary shelters, and perhaps it was the cold that made it all the more vital for humans to possess the flexibility and inventiveness afforded by a larger brain. In fact, a couple hundred thousand years before the oldest date in Cann, Wilson, and Stoneking's range for the mitochondrial appearance of Eve— right about the time that *erectus* seemed to be lightening up and assuming small but telltale physical traits that later would be used to define the taxon *Homo sapiens*—there were rather sensational suggestions in Europe that either advanced forms of *erectus* or archaic *sapiens* were hunting big, dangerous game and setting up impressive if not overly sanitary systems of housekeeping.

Nowhere were such practices more notable than in Spain, where a Madrid aristocrat, the marqués de Cerralbo, in 1907 began excavating a site at Torralba that yielded stone and bone tools, fragments of wood which apparently had been worked by humans, and the remains of *Elephas antiquus,* an extinct elephant with curved tusks ten feet long. These animals, nearly fifty of them, were also found a mile to the north at a village called Ambrona. They suggested that advancing humans had butchered the large mammals and piled up the bones.

The existence of these bones as well as indications of fire led fanciful writers and anthropologists to conclude, rightly or wrongly, that humans 500,000 years ago had used fire to drive the elephants into a marshy, miring area where they could be immobilized and more easily killed.

"That is probably the biggest controversy in the Middle Pleistocene of Eurasia," says Clark Howell, who has investigated both sites. "The press made much more of human hunting, driving, massive manipulation of fauna than we ever thought. It was not like North American Indian bison drives. But they were taking animals, they were perhaps bogging them in, in some situations, or exploiting them if they were already dead or dying. The breakage of the bone especially at Torralba has a pattern to it. Usually the bones are disarticulated, and we consider there's good evidence of breakage and utilization of bones. And there is fire, carbon and charcoal in patches in all occupation occurrences. There's no question there's lots of fire."

The bones appeared to have been flaked by stones, perhaps to produce picks and cleavers, and some believed there was also an indication that these humans may have used spears. Pointed hollows were found in the ground. Meanwhile, somewhere between 220,000 and 400,000 years ago near what is now Nice, France, there appeared to have been a band of humans who set up a camp on a sand dune shielded by a limestone cliff at the mouth of a valley. It seemed to be a temporary hunting camp, perhaps established and revisited as humans followed seasonal game herds. What was left behind in addition to tools were indications that this band had built a hut held up by posts that afforded a forty-by-twenty-foot living area.

There was a hearth. There may have been a windscreen to protect the fire. In all likelihood the humans slept near that fire, and a few steps away was what appeared to be a work area. A flat stone may have been the stool. And there were the telltale flakes—good old stone tools. There were other huts found in the area that appeared to be oval in shape and sapling-walled.

They were nothing like the condominiums in current-day Nice—fossilized human excrement was discovered in at least one hut not far from the dining area—but they kept the northwest wind out and dinner seemed to have included oysters, mussels, and perhaps (only perhaps) caught fish. There were also the remains of wild boar and mountain goat. One of the tools looked like an awl used to pierce hides. And pieces of ocher suggested they may have begun to draw or decorate their bodies in symbolism and ritual (though the ocher also may have been a natural occurrence of iron oxide).

So maybe Eve had the first makeup. And she almost definitely cooked. There was meat to roast, oysters to shuck, and an imprint at the French site (known as Terra Amata) that looked as if it had been left by a bowl. There were even those who suggested that these humans filled containers with water and dropped in red-hot coals to boil their food.

Although that would have been a significant innovation, there was less evidence of advancing technology in the stone tools themselves. "The rate of artifactual change during this long time span appears to have been remarkably slow," notes anthropologist Richard G. Klein from the University of Chicago. "However, there is a tendency for later hand axes to be more refined than earlier ones, as well as the tendency for later assemblages to contain a greater variety of recognizable tool types."

Did the variety of implements serve as a template for language, as life grew increasingly complex and grunts were no longer specific enough?

Whatever the case, the Acheulian implements, identified first with *erectus* and now with the *erectus*-like successors, were being found across the entire breadth of Asia and Europe. According to V. P. Yakimov of the Akademia Nauk in Moscow, there were, by 1980, more than one hundred Acheulian locations discovered in the Soviet Union, situated in the southern regions of Moldavia, the Ukraine and Caucasian Republics, central Asia and perhaps in the Altai Mountains and the Far East as well. In a cave at Azykh was a man who looked like a transitional form between *erectus* and early Neandertals.

If the archaics were in fact transitional people on their way to early *sapiens*, that might be interpreted as showing gradual evolution, and gradual evolution would upset popular theories put forth recently by biologists such as Stephen Jay Gould of Harvard, who believes that evolution progressed not gradually but more in those fits and starts: the theory of "punctuated equilibrium." What did he think of these seemingly transitional beings? "I don't think anyone knows what they really are," says Gould. "I hear all sorts of different reports about them, and those who are committed to the notion of intermediacy see them as right in between and others see them as essentially modern. I don't know: when Piltdown was found it was reconstructed as intermediate, but of course that was a fraud made of a modern braincase—so strong was the desire to see an intermediate."

It was difficult to prove just what kind of people were using what kind of tools. In neither Spain nor France had associated human fossils been found. But archaic *Homo sapiens* of around the same date as Ambrona and Terra Amata *had* of course been educed in other European locales—especially if the Mauer jaw was as old as some folks said—and in England was a famous fossil found during the 1930s in the Thames River Valley near the village of Swanscombe. It was believed to date between 300,000 and 200,000 years—perhaps existing between ice ages—and from the looks of its parietals and occipital it was fairly "modern." But at around that same time a skull of similar age and similar fame was discovered in Steinheim, Germany, and it had features, as did a skull from a place called Biache in France, suggestive of the earliest stage of Neandertal, which as I've said is also a subspecies of *Homo sapiens* but certainly not of the modern *sapiens sapiens* type.

"There is good evidence for an early appearance of what we should call *Homo sapiens*—or something distinct from *erectus* anyway—in Europe, also in Africa," is the way G. Philip Rightmire from the State University of New York at Binghamton sums it up. "Neandertals of course came around somewhat later and were primarily a European and southwest Asian phenomenon. The earliest of them probably aren't more than 100,000 years or so in age, whereas these individuals from Arago and Petralona and from Africa are a good deal older. Toward the close of the Middle Pleistocene, there are signs that some of these traits begin to change more rapidly. It is during this period that populations of *Homo erectus* must have given way to the first representatives of *Homo sapiens*."

14

WE WERE GETTING to the nitty-gritty, and soon the search would be narrowed to a mere handful of beckoning caves. According to Berkeley, Eve had nurtured her young south of the Sahara, and one way or another they had spread from there to swipe the entire Old World from the *erectus* descendants who had been entrenched throughout Europe and Asia for half a million years.

You could take all the burdensome Chinese and European names from the previous two chapters—that list of archaics that looked like an index for an atlas—and toss them in a file that said "IRRELEVANT." Anything that wasn't from Africa, it seemed, wasn't one of our ancestors.

Obsolete.

Counting for nothing.

Or nearly nothing. This was the Berkeley breakthrough. Everyone knew or thought they knew about the origins of pre-men like *Australopithecus* and *habilis,* but never before had anyone been able to sort through all that, to penetrate deep enough into human kinship that grew increasingly complex through *erectus* and the archaics, and come up with a solution for where and when the first modern men and women appeared.

It wasn't just big news but a rewriting of the oldest history books. Relating the human samples to one another by linking female lineages based on the individual mutations they shared, labeling them as to types, and then designating them to larger clusters—counting those mutations that seem to have ticked through the ages like a molecular clock—the geneticists had pinpointed a female who appeared to be the sole common link in our species.

She was the woman Becky Cann described as "a nameless, faceless great-grandmother 200,000 years ago." But she wasn't nameless anymore, not since they had begun referring to her as Eve, and precisely where she had lived was greatly narrowed down once two thirds of the Old World—that 21 million square miles known as Europe and Asia—was tossed out as the origin of anatomically modern ancestry.

The problem of how so many Eurasian *erectus* and their archaic offspring could totally vanish after the arrival of Eve's clan may have seemed a fairly daunting question to most people, but Allan Wilson described it simply as "a fly in the ointment." Was the ointment snake oil? No, he would answer, it was genetics like genetics had never before been applied, and anyway Wilson was already working on a solution to the question of what advantage the invading Africans possessed in overcoming all those ape-men.

Meanwhile his lab had already filled big voids in a paleontological record that Cann had found "frustratingly silent."

Although Africa had been an increasing likelihood for man's origins since the Taung child and the Leakeys' *Zinj*, the actual evolution into real people—real, anatomically modern humans, not just big-browed leftovers from *erectus* favored with a few modern traits—could just as likely have occurred in Asia and Europe, since they too were abounding in the rootstock archaic *sapiens* form. Right into the 1970s, Africa was considered fine for the birthplace of man-apes and other of humanity's oldest ancestry, but not necessarily for modern man himself. To many it seemed like a cultural backwater.

"Whether the transition to more modern humans occurred in Africa or elsewhere cannot be determined with present information," Rightmire himself emphasized as recently as 1981.

Now it could. The answer was in base-pairs of mitochondrial DNA. Rising from the savannah while those older humans spent all their time skirting the glaciers to the north, Eve's descendants had headed up to Morocco and the Middle East, fairly

brushing by more primitive people like the Neandertals, who represented the last archaic form, or stepping on their *erectus*-like
bones as they swept into Eurasia. It was an incredible scenario, the
thought that *erectus* and its progeny—themselves evolving, or
seeming to evolve, into more modern types throughout Europe,
Asia, and the Middle East—would suddenly and essentially find
their time-honored ilk totally replaced by immigrant Africans
who'd evolved faster out of the archaic mold.

Or who had some kind of mysterious advantage.

The issue was settled once and for all. No more fretting
necessary, no further need to check out more teeth in Vergranne
or La Chaise or La grotte du Prince. No more use salivating over
finds in Java or North Wales or the Upper Cave at Dragon Bone
Hill.

It was Africa.

The continent which had brought you the first fox-sized
apes and then *afarensis* and the robust man-apes and "1813" and
"1470," which had sent forth *erectus* as far as Java—now that same
continent was bringing out a whole new line called anatomically
modern humans to replace the far-flung ape-men.

Africa, it seemed, was like a giant car company that issued
one model of sedan but replaced it a while later with a wholly new
model. Yes, as many had long come to suspect, that landmass
second in size only to Asia, in large part out of sight beneath the
equator, was plainly and simply the fountain of creation. But also
the fountain of renewal, regeneration, and permanent change.
Was it any wonder that Cann called Africa the "cradle of human
polymorphism"? Something dynamic seemed to occur down there,
something spawning advantageous mutants, and from Africa these
products of periodic transformation radiated outward like waves.

And why fuss over that? There was no big problem in
saying Eve was from Africa. Not only was there an unassailably
lengthy history of apes, man-apes, *habilis,* and *erectus* there, but the
deep and dark continent had also been very rich in the kind of
transitional archaics who had to have spawned anatomically modern humans. This most people agreed upon: that anatomically
modern *Homo sapiens sapiens* had logically been born of archaic
sapiens. And in Africa, the archaics had been more than ade-

quately represented by Bodo, Ndutu, and Rhodesian Man. Clearly Africa had archaics to spare.

But so did the other two continents—Swanscombe, Dali, Petralona—and the idea of a certain select, "isolated" population of sub-Saharan archaics turning into moderns and spreading suddenly from the savannah to conquer all the world struck a lot of people as less than polite. And a little strange. It was a major sticking point. What Berkeley was saying, after calculating the diversity of mutations, after comparing every person with every other, after concocting a beautifully convoluted tree, was that a peculiar band of Africans had overwhelmed, had pushed into utter extinction, all those *erectus* and post-*erectus* archaics who had managed to survive just fine for such a very long time, who in the case of Java Man may have gone back close to and perhaps more than a million years, all those fledgling humans who had been entrenched from Java to Kwangtung, from Russia to the French Pyrenees, who had in many cases seemed to be shrugging off thick skulls and oversized jaws, whose glenoids foresaw an era of modernity, who had found a way around the glaciers, all had—simply disappeared.

Complete replacement. No admixture.

Gone like that, without a genetic trace.

No interbreeding.

Or nearly none. While they were conquering the world, Eve's grandkids didn't even have the decency to abduct a few women. That's what it came down to, or at least that was the impression that was given. That's what caused the uproar. *No admixture.* Not a single detectable Asian *erectus* gene. Surveying Asian mitochondria, the Berkeley group had found no DNA types that had a level of variability as great as the African samples, which indicated the Asian DNA type hadn't been changing for as long a period of time. Simple as that. With a dash of the pen, Wilson's lab was perceived to have negated the work of everyone who had pursued our ancestry anywhere outside of Africa. "Thus we propose," to repeat the *Nature* paper, "that Homo erectus in Asia was replaced without much mixing with the invading Homo sapiens from Africa."

"Without much mixing" left a little loophole for some

admixture. An *erectus* woman may have been abducted after all. In the *Nature* paper, Cann, et al., had inserted a few qualified verbs or adjectives (*"might* imply" or *"unlikely* that Asian *erectus* was ancestral to *Homo sapiens"*) but the implication was more forceful than that, and they ended up saying outright that while archaic types of mitochondrial DNA could have been lost from a hybridizing population, the probability of that happening was low.

In a paper Cann and Stoneking presented during a symposium at Cambridge on July 8, 1987, shortly after the *Nature* paper was published, they emphasized the possibility that they just hadn't detected a non-African lineage yet, or that mitochondrial DNA types from Asia were lost by selective mechanisms. But they also said "the frequency of unobserved types in this sample must be less than 0.3 percent," and if the two possibilities were eventually erased by further research, "the rather staggering implication is that the dispersing African population replaced the non-African resident populations without any interbreeding."

"Now what happened when those modern humans migrating from Africa met with the resident population, the descendants of Java Man and Peking Man and so forth, the descendants of the migration of *erectus* out of Africa a million years ago, I'm not prepared to say at this time because, although the observation is that they did not contribute any mitochondrial DNA types to modern human populations, there's at least three explanations you can think of for that," explains Stoneking, repeating his Cambridge qualifiers in a tone of caution that was not as obvious in the *Nature* paper.

"First it could be that we don't see any of these so-called ancient, non-African mitochondrial DNA types because we simply haven't looked in the right populations yet. There's still a lot of populations that haven't been sampled. Second, it could be that these populations did contribute mitochondrial DNA types to the colonizing population—the migrating populations out of Africa— but that their mitochondrial DNA types were subsequently lost, subsequently went extinct. This could be because they were at a selective disadvantage—invading populations had some sort of advantage in their mitochondrial DNA—or just went extinct at random. Any time a female either leaves no offspring or leaves only male offspring, her mitochondrial DNA type goes extinct. The third explanation would be that you don't see any extremely divergent, non-African mitochondrial DNA because they were never contributed in the first place: that those populations were completely replaced."

Whatever the nuances of qualification, Cann, Stoneking, and Wilson were placing their data in absolutely direct opposition to the fossil scientists who believed that evolution into modern humans probably occurred gradually, as genes got swapped back and forth in and between any number of places where the archaics were. As Cann, et al., phrased it in the *Nature* paper, "An alternative view of human evolution rests on evidence that *Homo* has been present in Asia as well as Africa for at least one million years and holds that the transformation of archaic to anatomically modern humans occurred in parallel in different parts of the Old World. This hypothesis leads us to expect genetic differences of great antiquity within widely separated parts of the modern pool of mtDNAs. It is hard to reconcile the mtDNA results with this hypothesis."

Whether or not Weidenreich was turning in the tomb, it was more than enough to turn any number of living paleoanthropologists gray. Not just Peking Man but Swanscombe and Petralona and those house builders near Nice, along with the Neandertals who came after them, the cavemen of classic literature, all were rendered into instant and total oblivion. The concept was radical, shades of old-fashioned free-speech days, as far out as a Yippie with a bong on University Avenue.

That's right: scotch Neandertals too. They were just a later sort of archaics: *Homo sapiens neanderthalensis.* And as was the case with *erectus* and Asians, "modern Europeans show no evidence in their mtDNA of ancient maternal lineages contributed by [Neandertals]." Their existence, starting perhaps 125,000 years ago, overlapped with the later stages of the archaics, and they were sprouting up from the archaics in Europe while Eve's more modern kind were evolving from the archaics in Africa. At the same time Eve's descendants were preparing for the proliferation from some isolated sub-Saharan archaic locality, the Neandertals were making an unenviable living from Portugal and especially France

Neandertal

to territories inside of what is now the Soviet Union. So striking were certain Neandertal samples from Russia, especially one cited from a place called Teshik-Tash, that one writer, V. P. Yakimov from the Institute and Museum of Anthropology in Moscow, voiced his feeling that the USSR's southern territories—such as the Caucasus—"could have been a part of the ancient homeland of *H. sapiens.*"

But it didn't appear to have worked out that way. Alas, Neandertals had the misfortune of being all over Europe, shambling from cave to filthy, breccia cave, when Eve's troops came marching in, disqualifying both them and any other type of remaining archaics.

Well, really not *instant* oblivion. It wasn't quite as radical as some people made it sound. And if there might have been a bad case of myopia, the logic was often impeccable. Cann figured the elimination of more primitive humans by Eve's African offspring may have taken 100,000 years. In the course of those thousand centuries, Eve's progeny, and the progeny of others with whom she'd grown up, had swiftly assumed the physiological attributes of modern man, and somehow they were better equipped— uniquely equipped, it seems—to survive to the present day. It was not necessarily a case of archaic warfare, but much more likely, Cann stressed, a case of outcompeting.

What was so unbelievable or myopic about the hypothesis when it was phrased that way?

Although it couldn't have been brain size—the older archaics were within the range of modern humans, and Neandertals had brains that were bigger than ours—Eve's children had some kind of inherent advantage over the humans in Europe and Asia, and they were so adept at exploiting the environment that they presumably left few resources for anyone else. Over those thousand centuries, Eve's kind kept completely away from interacting, at least reproductively, with more primitive *sapiens,* who must have stared awestruck from their caves as these lithe new wunderkind spread determinedly through the terrain, not only to push them out of Eurasia but also to migrate into Australia and the Americas.

The march of *Homo sapiens sapiens.*

While the range of time they'd presented for the actual

migration out of Africa was a wide range indeed (23,000 to 180,000 years ago), and while Cann and Stoneking further hedged their bets in the Cambridge paper, now placing Eve in sub-Saharan Africa 50,000 to 500,000 years ago, which looked like a little backpedaling, it seemed most likely that her descendants had left sometime around 90,000 years ago and that by 40,000 years ago *Homo sapiens sapiens* had completely replaced *erectus*-like *sapiens*. And the modern men from Africa, by this time too, had found their way to Australia and soon after that New Guinea.

Very interesting was the Berkeley group's discovery that each region under study was colonized not once but repeatedly. There were any number of maternal lines. For instance, it seemed that members of not just one but fifteen lineages had found a way to cross the ocean from Asia to Australia, because the Aborigines seemed to descend, explained Cann, from that many branches on the DNA tree. That went contrary to conventional wisdom, which saw Australia founded by some lucky adventurers setting out blindly from island to island in a big canoe. According to Berkeley, the founding of Australia was instead by a people who knew exactly where they were going and had gone there time and again. It also meant that those first settlers were not the archaics from Indonesia that Weidenreich and Dubois had thought looked like "proto-Australians," but instead had been Africans. Asia's current populations came not from Asian archaics—not from Ngandong Man—so much as from archaic Africans.

About 12,000 years ago, some of these African *Homo sapiens sapiens* who had settled in Asia and had evolved into the race known today as "Mongoloid" expanded upward and across a landbridge between the Soviet Union and Alaska, establishing the American Indian populations.

Cann bolstered her, Stoneking's, and Wilson's hypothesis with a little archeology. She pointed out that Africans were using sophisticated blade tools, "meticulously manufactured from fine-grained rock," about 90,000 years ago, while the hapless Europeans and Asians, the archaics who soon were to find themselves out of mitochondrial luck, were still using flake tools "made less carefully from whatever rock was at hand."

Stoneking wasn't sure what remarkable advantage the invading *sapiens* must have had. "People have pointed to various things, like how different invading populations may have been physically," he told me. "Were they even capable of interbreeding, as well as what one could imagine in the way of technological or cultural advantages a population may have had: better weapons and so forth?"

Cann took stronger stands. She fairly scoffed at those who believed the archaics had evolved into *Homo sapiens sapiens* in more than one place at a time. "Most of the anthropologists reject this theory because of the great similarities in skeletal and cranial shapes that people throughout the world exhibit today," she said, emphasizing that if evolution into anatomically modern humans had occurred at more than one place, there would be more than one distinct form of present-day human. "Thus, the preponderance of paleontological evidence supports—though it does not prove—the theory that modern humans evolved in Africa and then emigrated to Europe and Asia. . . ."

Moreover, Cann had said that the first modern humans did not simply migrate somewhere and dig in to form a new race. Had that been the case, the largest limbs radiating from the most populous branch on their tree would have contained a single race. Each race would have had a distinctive mitochondrial DNA. Instead, each limb except the solely African one held a whole assortment of mankind, and that implied that the various races—black, yellow, white—had established themselves only after the modern African invaders had moved back and forth around the world several times.

Only then had they settled down in sufficient isolation to develop racial characteristics. Each limb on the mitochondrial tree, except the African one, showed an assortment of races. As Cann pointed out, one New Guinea type had as its nearest relative an Asian. Other New Guineans appeared to be closer to Africans than Australians, even though these two islands were connected at various times in the past by a bridge of land and the natives bore so many resemblances that some anthropologists had considered them part of one big Pacific population.

This is the sort of thing Ernst Mayr had found "staggering." Mayr used a different example, saying that placing the African population "in a different category almost from all the rest of the world" indicated that "for instance the western Europeans—the straight Caucasians—are more closely related to the Australian Aborigines than they are to the African black."

Yet it fit perfectly well with the hypothesis: recent racial differentiation. After all, Cann had said that early modern humans moved "back and forth around the world" several times, intermixing with each other, before settling down "in sufficient isolation" to develop racial features.

"That one branch holds only Africans indicates that some of the original *Homo sapiens sapiens* remained in Africa," she wrote in *The Sciences.* "But thirteen of the twenty Africans in the survey are scattered along the branch that includes the four non-African races, which proves that some populations migrated out of Africa, mixed with European and Asian populations, and then returned."

Modern humans from Africa moving all over the Old World and into Australia and New Guinea too, then crystallizing into a situation of enough isolation so that suddenly racial differences took root—that is what the hypothesis amounted to. After a lot of interbreeding among modern humans once archaics were out of the way, bingo: isolation and raciation. This made it possible for an Aborigine to be closer to a European than to a black African.

In no time at all, three letters had materialized in *Nature* challenging the Eve findings, the first just months after publication, on April 23, 1987. It was from Robert B. Eckhardt of Pennsylvania State University, who pointed out that there were skeletal characteristics that seemed to link modern Asians to the old Asian *erectus* who supposedly had been shoved into oblivion—rendered genetically irrelevant, or close to genetically irrelevant, removed as a meaningful ancestor—by the onslaught of Eve's descendants.

How could modern Asians have traits associated with Asian *erectus* if the Asian *erectus* had been wiped out?

This was the crux of one upcoming debate. Moaned Eckhardt: "there is insufficient cause to discard the first half million years or so of Asia's hominid-fossil record."

A month later there was another letter, this one submitted

by Naruya Saitou and Keiichi Omoto of the University of Tokyo, who stated flatly that "data from mitochondrial DNA, which can be regarded as a single locus, are not enough by themselves to establish the branching pattern of human populations."

The following September two Paris scientists, Pierre Darlu and Pascal Tassy, echoed the complaint that the mitochondrial DNA was only a small part of a person's total genetic material ("being .048% of the whole human genome"), argued that Berkeley's genealogical tree, which used the method known as "midpoint rooting," might not give the same tree as one rooted by an "out-group" that was known to be an actual ancestor (such as a chimpanzee), and raised the especially sensitive issue of Cann, et al., using American blacks as their "African" sample. The tenacious Berkeley group countered with a response and argued successfully against publication of another critical letter *Nature* had considered.

And they traveled around to seminars presenting the theory and being greeted mainly by baffled silence or expressions of subdued wonderment—at first, anyway, while everyone was trying to digest what they were claiming.

"Shortly after this came out, I went to the meeting in Cambridge of anthropologists and found that there were a few who were real excited, real enthusiastic about it, but they were the ones who generally favored an African origin anyway based on their interpretation of the fossil and archeological record," says Stoneking in the kind of matter-of-fact tone of a weather forecaster who is predicting another dry day during a summer-long drought. "There were a few people who were critical to the point of being irrational—they tended to be those who did not favor an African origin—and most people were sort of in between, thinking it was interesting but not really understanding it."

It was difficult to understand mainly because it was phrased in a foreign language, the language of genetics, kind of like someone bursting into a room, talking with animation and great conviction in Greek, and then asking an English-speaking audience what they thought of what had just been said. And expecting everyone to understand a lingo familiar in some cases only to those who'd actually worked in the lab. Wilson discounted the paleoanthropologists before they really had a chance to react. He pointed out that back in the 1960s, when he and Sarich had said the first hominid evolved 5 million years ago (and not the commonly thought 15 to 30 million years), they were initially ridiculed and it had taken "a decade and a half before the anthropologists realized they were wrong." Now he said smugly: "Some people don't like our conclusions but I expect they will be proved wrong again."

Actually, the reaction was not so negative at first. Harvard's Howells didn't have much of a problem with it, the Leakeys liked the new technology (and doubtless the fact that it supported an African origin, though Mary snapped that the term "Eve" was "nonsense"), Pilbeam had no searing objections, showing little lingering bitterness over how previous molecular data had trashed his *Ramapithecus* theories, and Mayr was puzzled over some of the implications but not in opposition, acting rather laid back. English paleoanthropologists such as Chris Stringer of the British Museum of Natural History were extraordinarily enthusiastic and Stephen Jay Gould seemed nothing short of in love with the Eve proposition (though he didn't want to say much about it "because of the ugliness of the debate" that was about to break out). Clark Howell too was "overjoyed." Rightmire saw it supporting views he had himself long held, and if Lucy discoverer Donald Johanson didn't buy the mitochondrial theory, his close colleague Tim White did.

On the other hand, Phillip Tobias in South Africa didn't think a whole evolutionary theory could be based on such a small aspect of genetic material; Alan Mann saw the theory as increasingly difficult to maintain as more and more data was analyzed; Randall Susman of Stony Brook felt the way Wilson played the scientific stage hurt the field's credibility; the University of Tennessee's Fred H. Smith thought the DNA conclusions "dogmatic"; C. Loring Brace, a leading anthropologist and textbook author from the University of Michigan, believed there was "no reason to sustain" what he called an "incredible" theory; Elwyn Simons saw the Wilson lab's announcements as scientific entertainment; and Wilson's two closest allies from the 1960s, Berkeley's own Sarich and Sherwood Washburn, maintained their serious reservations.

But weighing in with the most powerful opposition was Michigan paleoanthropologist Milford Wolpoff, who more than anyone else was upset about the Asian *erectus* being tossed away. He had taken it upon himself to "turn the tables" on the Berkeley group. *The closer you look, the messier it gets.* And soon he wasn't just turning the tables, he was overturning them.

15

Milford h. wolpoff was often described as the man who had examined more fossils than anyone alive. He didn't actually dig for artifacts. Rather, his expertise was measuring, interpreting, and comparing what fossil hunters already had at hand. He had seen it all, from modern populations in Australia to the oldest *Australopithecus.* He knew the European archaics well enough to write whole papers on them, and he was one of the world's foremost experts on the *erectus* from Asia. When it came to analyzing fossils, his name was as frequent in some journals as the Leakeys were for the discoveries themselves. Whether or not anyone could lay claim to examining the most fossils, Wolpoff was the walking personification of paleoanthropology, and he *detested* DNA.

That is, he strongly disliked how Berkeley was using the stuff, and he hated what he saw as overblown claims. He did not believe there was any such thing as a molecular "clock," arguing against it since the old blood protein days. His was also that first voice of opposition in the *Chronicle* article on an "African Eve" back in 1986.

Wolpoff was a natural enemy of molecules, and when Cann, et al., referred to paleoanthropologists in the negative, as they had in those preemptive strikes early on, it was people like Wolpoff they had in mind. That had rankled him, things like Cann saying "it is too much to hope" that "the trickle of bones from fossil beds" would "provide a clear picture of human evolution any time soon." He might even agree to a certain extent (and by the way, he had never been a supporter of *Ramapithecus*), but it was still irritating when she kept twisting in the fact that "the traditional portrayal of fossil evidence" had left the impression that "scientists have reliably mapped the evolution of humans" when it was obvious, said Cann, that they hadn't. She, Stoneking, and Wilson had made it clear that they didn't think their critics understood the new technology.

They seemed awfully condescending. They were also bewildering. Cann would throw down the gauntlet and in the same breath extend an olive branch. "This could be the start of a wonderful working relationship," she had concluded in one paper—after declaring that *her* genes were superior to *their* fossils and making the provocative remark that museum dioramas contained "more fantasy than Spiderman's best escapades."

While paleoanthropologists could never be sure a certain fossil actually left descendants, she emphasized, there was "100 percent certainty that genes in modern populations have a history that can be examined and will trace back in absolute time to *real ancestors*" (author's emphasis). Through genes, Cann pointed out, they had found a female who was probably the only common link in our species, "and through those genes, we are uncovering areas of the past that have been hidden from us by history, culture, and our own eyes." That was rapping them over the head with the olive branch.

Wilson referred to Wolpoff and his colleagues as "certain loud voices raised in Michigan." He was still obsessed with their initial rejection of his albumin results. Now he was sure they'd be eating crow just as other anthropologists who'd dared to criticize him before had eaten crow over *Ramapithecus*. "The anthropologists said the first hominid evolved at least 15 million years ago, and we said it was more like 5 million," he reiterated to the journal *Science*. "It took a decade and a half before the anthropologists realized they were wrong." And now, in the wake of the new DNA "clock," to repeat what Wilson thought, "Some people don't like our conclusions but I expect they will be proved wrong again."

At the same time all those feelings were starting to run so high, Cann, et al., had not been able to avoid citing Wolpoff's work in their own *Nature* paper. Everywhere you went, there, in citation parentheses, or at the top as a co-author, or in the bibliography, was that name: Wolpoff. He had started out as an expert on australopithecines and now he was an expert on Neandertals and *erectus* and European archaics. He could argue obscure points of interpretation in hominids and humanoids scattered across three continents and 4 million years. He also had a working knowledge of genetics. And he had one word to describe what was coming out of Berkeley:

"Wacko!"

That was Milford Wolpoff, a character as well as a scholar, looking a bit like Ernest Borgnine, street smart, son of a Chicago taxi driver. For kicks he played the clarinet in the Ann Arbor Symphony Band. At conferences he held debates in the pubs. That was where some of the best discussions with fellow anthropologists occurred, sudsing it up after a long dry day of seminars. He was a very serious scholar, and a very thorough one. Everywhere you went: (Wolpoff). And though, like many of his colleagues, his ego was large enough to consume the western end of Olduvai Gorge (and many of his associates complained that a conversation with Wolpoff was usually a Wolpoff monologue), he could be a teddy bear of a man, very witty, generous with his knowledge, and his ego somehow didn't translate into offputting haughtiness.

Unless you were his opponent. Then his ego was just fierce and abrasive. Cann was clearly daunted by his experience with fossils. She had fairly shrunk from him one time when they met. "There was some worry on her part about whether I was going to chew her head off," says Wolpoff with a twinkle in his eyes. "But I didn't."

She thought him "real bombastic," and accused him of quoting her work out of context. At one convocation he had attacked her rather personally, and her allies like Chris Stringer at the British Museum had also borne Milford's wrath. "Historically the British Museum has been responsible for many important advances in paleoanthropology," said Wolpoff during another conference in Europe. Then he flashed a slide of the fraudulent Piltdown skull.

A teddy bear or a grizzly one? Call this chapter "Cann and

the Cave Bear": Wolpoff was going to go full force and claws bared after Berkeley, leaving no rock unturned, fighting even on genetic grounds, while off to the side, once everyone had begun to understand what Cann, et al., were up to, there were any number of other skirmishes breaking out.

Wolpoff thought the use of American blacks as a sample for Africans, to start off with, was "bizarre, ludicrous." The entire DNA hypothesis was untenable, in his opinion. For the DNA "clock" to function, he figured, the mutations had to be both constant and random, implying that this type of DNA was not distorted by forces of natural selection (whereby certain mutations are eliminated and thus would not be there to be counted centuries later by geneticists). In other words, the changes in mitochondrial DNA would virtually all have to be "neutral," which he didn't believe was the case. He also didn't believe for a minute that Africans could take over whole continents without interbreeding with the resident populations like *erectus* and Neandertals.

"I don't think Rebecca thought the consequences through, I think she just got carried away by believing she was right," he says. "It's like preaching a new religion. If fossils don't fit her model, she feels it's our problem. I had an adviser who was a geneticist, I took population genetics, I think I understand this stuff. And what I want to see is the model where one species of humans replaces another through only subtle competition without directly interacting with each other. And if they're not different species, wouldn't one assume that the women would be incorporated? Do they envision that the archaics and *Homo sapiens sapiens* lived next to each other or waved at each other as they walked by, but never mixed?"

If the Berkeley group was right, thought Wolpoff, then someone should be able to find African-like fossils in Asia from around the time of the hypothetical invasion. Yet no one had. "We've never found the combination of small browridges, high foreheads, and jutting faces of the Africans in Asia," he argues.

Instead of bearing African features, says Wolpoff, Chinese and other Asians today bear resemblances to those old *erectus* populations that Cann and her colleagues say went extinct. They did not go extinct, argued Wolpoff. Peking and Java Man instead contributed substantially to modern gene pools. He had the fossils to show it. And what they showed, those fossils, was a facial and forehead form that is still detectable in modern Chinese. In regions where the fossils had flat faces, one could find flat faces today. That didn't speak well for the concept of an African "invasion."

"It's a neat story, the out-of-Africa hypothesis, and if I was a graduate student I'd be sucked right into this," he said. "But there are problems. The problem is that as Eves left Africa, they left as Africans, but when they arrived in Asia, they arrived as Asians! When they arrived in Europe, they arrived as Europeans!"

In presentations at places such as the University of Pennsylvania, Wolpoff used more slides and more sarcasm to get across his point. One was a photograph taken from the rear of Africans walking down a path with baskets on their heads. That represented the migration from the sub-Sahara. In the next slide he showed a Chinese family smiling as they approached on bikes. *That* was Eve's clan arriving to Asia.

"Wacko!" repeated Wolpoff, determined to show just how ridiculous the Berkeley hypothesis was. It came down to this: if he could demonstrate genetic continuity somewhere outside of Africa—if just one place in Europe or Asia had connections between old cavemen and modern humans—he felt that would be enough to disprove the idea of a common recent origin and an African takeover. Evolution into anatomically modern humans had taken place across the board, he believed, with archaic populations in Asia, Africa, and Europe all turning modern as genes and cultural ideas flowed between them. This was called "multi-regional continuity." *Erectus* in Africa served as the founding population for current-day Africans, *erectus* descendants in Europe for current Europeans, and *erectus* in Asia for any number of populations ranging from Australians and Eskimos to Chinese and American Indians. There was no single and magical "Garden of Eden."

Although Wolpoff had modified it, the theory of continuity went back to Franz Weidenreich and Carleton Coon, two of the science's most historical figures. Like Wolpoff, who had assumed his mantle, Weidenreich was an indefatigable morphologist and an unambiguous man. He made clear his position that "there was certainly not one Adam and one Eve who could be claimed as a progenitor by every living man today." While Wolpoff disagreed with certain of their precepts and thought some of Coon's ideas had been rightly rejected as racist, he did not want to "throw the baby out with the bathwater," and the baby was that concept of regional continuity.

Both Weidenreich and Coon had written in detail of similarities between *erectus* populations in Asia and modern man—exactly the type of stuff that contradicted the Berkeley replacement theory. Weidenreich in fact was the first to see the supposedly unbroken chain of links from the Java *erectus* through the archaics of Ngandong and on to living Australian Aborigines. This in turn resurrected the memory of how Dubois himself had described the Ngandong archaics as "proto-Australians." There were also those resemblances between the fossils at Zhoukoudian, where Peking Man had been found, and modern Chinese. Take flat faces: they were present in both the fossils and current Asians. The inescapable implication was that whatever the role of Africa back in the days of man-apes, the origin of modern Asians was Asia.

Coon, a Harvard man who retired in 1963 as curator of ethnology and professor of anthropology at the University of Pennsylvania, had taken it upon himself to interpret much of Weidenreich's work and had listed a whole slew of features which he said were shared between the *erectus* from China and living Mongoloids. These common traits included a ridge down the middle of the skull found in North Chinese, bony growths of the mandible that were noticeable in 2 to 5 percent of Japanese, Lapps, and natives of Siberia, a general thickening of the tympanic (outer ear opening) plate found mainly in Eskimos, American Indians, and Icelanders, a special growth on the tympanic plate found in 20 percent of Polynesians, broad nasal bones, and incisor teeth that were oddly shovel-shaped in both living Asians and Peking Man. As for the caveman from Java, he'd had a skull ridge (or sagittal keeling) that appeared to have been handed down to modern Australians and Tasmanians.

This argued powerfully for regional continuity in Asia from ancient times to today and certainly suggested that the Asian *erectus* had not been rendered totally obsolete. The problem was that aside from tossing in certain half-baked notions with his morphological evaluations, Coon was astonishingly indifferent to racial sensitivities.

"My thesis is, in essence, that at the beginning of our record, over half a million years ago, man was a single species, *Homo erectus*, perhaps already divided into five geographic races or subspecies," Coon wrote in his classic book *The Origin of Races* (1973). "*Homo erectus* then gradually evolved into *Homo sapiens* at different times, as each subspecies, living in its own territory, passed a critical threshold from a more brutal to a more *sapient* state, by one genetic process or another."

That wasn't so horrible, but then Coon went on to make that statement about how Australian Aborigines were still in the process of "sloughing off" *erectus* characteristics (their browridges, as I said, were unusually pronounced) and included the photograph in a previous book of an Ituri Pygmy that was obviously intended to show how much the Pygmy's jutting jaw resembled a caveman's. Coon believed that after a certain point in evolution, major African populations had stagnated, lagging behind the other races.

If that wasn't enough to get the baby thrown out with the bathwater, Coon wrote this gem in a third book:

[Anthropologists might opt for the preservation and wit-sharpening of elderly Australian aborigines and Bushmen, to serve as permanent informants to future generations of students. . . . The negroes, meanwhile, have another innovation to look forward to. Recent research on the actions of two hormones secreted by the pineal body make it possible that before long people will be able to change their skin color whenever they like, by simple injections. A colored woman could thus turn white with less effort than it takes to have her hair straightened, waved, and set. This would be particularly effective for those with narrow features and dark skins.

That was strong stuff even for a Klansman in the Louisiana legislature. No wonder Wolpoff denounced Coon for "silly" and "racist" ideas. But there was the tradition of regional continuity buried in all that gunk, and despite everything else, Coon had been a brilliant morphologist. Many of his observations remained relevant. Wolpoff felt it his task to scrape Coon's crud off Weidenreich's old work and excavate veins of truth that had been buried along with Coon's various outrages.

In 1978, Wolpoff had gone to Indonesia and had reconstructed an *erectus* found on the island where Java Man had been discovered during the last century. The previously unreconstructed fossil, "Sangiran 17," retained a very vital feature missing in most such fossils: its facial skeleton. Wolpoff reassembled it himself, holding it on his lap and gluing the various pieces together. When it was dry enough to move he had looked at the skull from various angles, and to his astonishment the 500,000-year-old fossil looked like an Australian Aborigine. "I just never believed in continuity until I saw that skull from the side," he says with lingering amazement.

That characteristics of the skull and face could persist in one region over the course of half a million years did not bode well for Garden of Eden theories of population replacement with little or no interbreeding. More than anything, it was the impact of the entire suite of features that impressed the Michigan paleoanthropologist. When Wolpoff, trying to quantify this overwhelming impression, measured the skull and compared it to some Aborigines who had been exhumed from a series of graves at Kow Swamp in northern Victoria, Australia, he found that both the aboriginal specimens (which were at least 9,000 years old but good samples of anatomically modern men) and this *erectus* from Java shared a distinct flatness of the skull frontal, a rounding of the lower border near the eyes, similar facial height, and teeth that fell within the same dimensional range. Though a characteristic so general that it might have also been found in any number of populations elsewhere, the nasal breadths of the *erectus* and Aborigines served as a metaphor for the striking closeness of certain traits: 26.4 millimeters in Sangiran 17 and 26.8 in the Aborigines.

Nor was that the whole of it. There were also the Ngandong specimens, which seemed like intermediates between *erectus* and *Homo sapiens,* and there was another "intermediate" from a place called Sambungmachan, and there were what are now called the

Wajak skulls, an example of the earliest Java *Homo sapiens*. The picture forming in Wolpoff's mind was that of continuity in Indonesia from Java Man to Sambungmachan to Ngandong to Wajak to modern Aborigines. The Australians, who were thought to have originated in Indonesia, shared a number of features with archaics such as Ngandong Man, including long, flat, receding foreheads and indeed browridge similarities. Meanwhile, in Wolpoff's opinion, a 28,000-year-old skull found at Lake Garnpung in Australia (known as "WLH 50" for the Willandra Lakes region) was also like the far more ancient Ngandong Man, especially in its thick cranial vault, general robustness, and the shape of the frontal just above its eyes. "The Australian skull shares twelve character states uniquely with the Indonesians," wrote Wolpoff, while not one character state, he said, was uniquely shared with a similarly archaic African.

As for China, Wolpoff, like Coon and Weidenreich before him, saw, in the famous Zhoukoudian caves, a veritable time capsule of continuity, from Peking Man to modern *Homo sapiens* dated at 15,000 years in the Upper Cave. Anything dated so recently as 15,000 would have come long after Berkeley's alleged replacement of Asian *erectus* by those sub-Saharan Africans, and so should not have resembled an Asian *erectus* whatsoever. Certainly, if *erectus* had left no genes, none of its unique characteristics should show up in present-day populations; and yet, in addition to the flat faces and shovel-shaped incisors that Chinese ape-men possessed long before Eve's time and which are still prevalent in Mongoloid populations today, there was a Chinese mandible of more than half a million years that showed the absence of third molars, a condition not uncommon in modern Chinese. The archaic from Dali, meanwhile, possessed a naso-malar angle of 145 degrees, approximating the flat facial angle of Eskimos, who are descendants of Asians. In the Philippines yet another cave yielded browridges reminiscent of Zhoukoudian, bringing still more of Asia into play.

Working assiduously with Alan G. Thorne, a prehistorian from Australian National University in Canberra, and Wu Xin Zhi, a paleoanthropologist in Beijing, Wolpoff had set about propounding a new view of regional continuity, and in 1984, just three years before the Eve hypothesis, they co-authored a 61-page treatise on fossil evidence from East Asia, replete with both extensive personal analyses of fossils and 274 source citations. In it, Wolpoff envisioned the human race as remaining ethnically distinguishable because populations had existed in certain regions long enough to acquire distinctive local traits, all the while maintaining an evolutionary course in the same general direction—remaining all one species worldwide—because there had been genetic contact (or "gene flow") between the various populations.

That gave much more a sense of brotherhood, equality, and one human kind. But it had no room whatsoever for a single point of recent modern origin, as Berkeley was claiming. "I can go on forever," said Wolpoff, when I paid a visit to his home in rural Michigan. "The fact is that there are features like this that show continuity in China, and what they seem to show is that there are no invading African populations, no place where different African morphology is intrusive. Each region of that part of the world has different features of what is primitive. There are faces that look Chinese even though they're found on vaults that look like *Homo erectus*. Dali is an example. It happens in every region of the world. I call this 'multi-regional continuity' and emphasize gene flow more than anyone has before. I didn't think gene flow was so important earlier in my career but now I believe it's fundamental to all change. Changes don't happen in local isolated populations, but rather they happen in populations connected to each other."

Wolpoff agreed in part with Richard Leakey, who in his book *Origins* (1982) saw the transition to modern humanity as like taking a handful of pebbles and flinging them into a pool of water: "Each pebble generates outward-spreading ripples that sooner or later meet the oncoming ripples set in motion by other pebbles. The pool represents the Old World with its basic *sapiens* population; the place where each pebble lands is a point of transition to *Homo sapiens sapiens*. . . ."

Becky Cann responded that the idea of continuity was an "outdated" one. Referring to the conception that "large-bodied, big-brained, long-lived animals spread out over three continents suddenly converge on the same genetic basis of whatever kind of morphology and intelligence and become one big species," she noted bluntly that "speciation doesn't work that way. It occurs in small, isolated populations and moves them divergently. So I think it's unlikely that the model of continuous evolution is correct for humans."

As for the incisor argument and other similarities between the Asian *erectus* and Asians of today, Cann implied that such seemingly unique traits could have evolved two different times in two separate populations because of similar environmental factors that spurred a peculiar adaptation. In other words, while the Chinese *erectus* had shoveled incisors, maybe some of the invading Africans had or later developed that trait too. Anyway it was quite clear from the DNA results, she wrote, that "the roots of Asian mitochondrial DNA diversity are simply not as ancient as those in Africa. Significant genetic contributions of people from this ancient Asian line of descent should have produced modern Asian maternal lineages which at least match the age of the African ones. Since they did not, we conclude that even though fossils of considerable antiquity are known from Asia, they were replaced by lines of African origin."

Wolpoff liked Cann but resented how the geneticists had an "arrogance that they're trained in a 'harder' science than we are, and that there's something more scientific about work done in a laboratory and on a computer than work done over bones in a museum. It's a combination of arrogance and that geneticists are not familiar with the fossil record. When you say the word 'arrogance,' Allan Wilson comes to mind. I sometimes get the feeling that what the geneticists do is read through our literature until they find something they want to hear, and then quote it, all the while sneering at paleoanthropology!"

But where were the geneticists going to turn once they admitted that their date was wrong? "Maybe to the fossil record!" says Wolpoff with glee.

And his voice was becoming more than just distant thunder. He was gathering his troops. It was clearly an urgent effort. He had to halt any more defections to the Garden of Eden school. Already the followers of Weidenreich, according to Howells, were probably a minority.

Wilson tried to ignore Wolpoff, showing his usual disdain. He closeted himself in his office out of reach of critics, disliking conversation with people who didn't have a biochemistry degree. When I asked him about Wolpoff, Wilson replied simply: "I persist in trying to distinguish between fact and opinion and treat the data in a statistical framework."

Like Cann, Wilson had less than high regard for fossil interpretations. "I know from previous experience that they can feel quite confident that they're right about something when in fact they're wrong." While it was true that paleoanthropology was more often than not in utter disarray (Wolpoff himself freely acknowledged that paleoanthropologists often tried "to pull something out of what isn't there"), and while it was also true at least initially that mitochondrial DNA was difficult for an outsider to comprehend—complex, its language an inbred foreign language, heavy on logarithms, unclear charts, and computer jargon—Wilson reacted with what had the appearance of simple intolerance. When questions came that didn't reflect wholesale acceptance of his results, his immediate reaction was that the skeptical questioner didn't have a "feel" for the subject, and he was too busy to explain it.

"Wilson at times was much less arrogant than now," says Sarich, no shrinking violet himself. (After all, Sarich was known to describe his own thinking as "brilliant.") But Wilson didn't have the charm Sarich did, and he seemed to believe that the nickname for his MacArthur grant was an accurate way of putting things. That nickname of course was the "genius award."

Big egos here, all trying to tell everyone who their grandmother was. But with Wilson behind the barracks, and Sarich only indirectly involved with the mitochondrial DNA, the task of fending off Wolpoff, and bearing the brunt of his quickly materializing offensive, had fallen upon Cann. She stood at the front line as Wolpoff began churning out prodigiously detailed arguments on how the fossil record proved Berkeley wrong. They had to be wrong. This "wacko" Eve hypothesis was in irreconcilable opposition to his cherished theories of evolutionary continuity and regionalism. Did they think he'd spent all that time in Jogjakarta for nothing?

And how was it that all the *erectus*, archaics, and Neandertals from Eurasia simply disappeared? No one from Berkeley had convincingly answered that yet. To a number of paleoanthropologists, there was only one way for the Eve hypothesis to have happened, and that was by means of some kind of holocaust or warfare. "This rendering of modern populational dispersals is a story of 'making war and not love,' and if true its implications are not pleasant," wrote Wolpoff.

It was a concept that Wolpoff used to irritate the Berkeley group, and it was a concept that made Cann want to scream.

16

"KILLER AFRICANS" is what Wolpoff started calling them. If it was true that Eve's descendants had taken over the world with just about no interbreeding, somehow causing archaic Europeans and Asians to disappear, then Eve's descendants must have been homicidal—they must have killed everybody off—and the hypothesis should be renamed after Eve's son, Cain.

"I like to write about their Pleistocene holocaust, because that's what it is, according to their model: a group of murderers from somewhere who killed everyone else off," insists Wolpoff. "I don't believe this, but they explicitly or implicitly must believe this, or there's no data for them to account for. If there is a recent origin to modern people, that means there's a supposition that the Berkeley group is making that there was no mixture. So they have supposed, whether explicitly or implicitly, that this replacement was complete. A complete takeover implies a holocaust of some sort."

"That's really thinking," shot back Cann, "at a caveman mentality."

That is, Wolpoff was throwing everything back to the outdated image of savage, prehistoric people, she complained. He knew it, too. It was how he gave it back to the geneticists, creating a straw man. Wolpoff knew it'd make Cann squirm. She detested such a notion and there was no evidence for it: that the conquering of the world by Eve's offspring was a violent conquering—it was outrageous. Although there'd been savagery among *erectus*, that was before the wondrous transition to *Homo sapiens sapiens*, and a long way from world war. Wolpoff was backing them into a corner from which there would be no easy escape. Not even Berkeley could prove a "holocaust," yet if populations didn't interbreed, didn't interact in some meaningful fashion, what other way to explain it than that they were separate species and fought each other to the death?

It made the normally smiling Cann inject a distinct sternness into her voice. "Wolpoff has immense experience, and he's also been very *wrong*. Evolution works by *out-reproducing*. All you have to do is last—offer more offspring. You don't have to clobber anyone over the head!"

She'd never said, never written, never thought such a thing. Holocaust!

"Wolpoff is a master of quoting out of context."

Well, she did use the word "supplant" at least once in her papers, and the secondary definition of that word, in Webster's, is "to supersede, especially through force." But that's not what she'd meant. Holocaust! It certainly hadn't been in the *Nature* paper, and she had dismissed such an idea in a flurry of articles she published during 1987 in somewhat less technical magazines like *The Sciences*. "Cultural innovations allow groups to dominate and replace even well-adapted indigenous populations," she'd emphasized in one of the pieces. It was a matter of a small population, spawned by one of the African archaic groups, producing more viable offspring than the primitive, entrenched people, and knowing how to utilize the environment to a greater degree.

Perhaps Eve's stock had a better ability to remember and plan ahead. Perhaps they were more organized. They probably made better stone tools, and as they prospered and multiplied, they may have consumed more than the usual share of fruit and game—causing the archaics sharing the same terrain to suffer shortages. Their lithe bodies may have conserved energy, since it was tough lugging around a bulky torso like the archaics had done. They probably built better hearths for warmth against the cold, too. And the *Homo sapiens sapiens* may have been more adept at treating ailments—or avoiding accidents—that were fatal to more archaic people.

According to Dr. Ezra Zubrow, an anthropologist at the State University of New York at Buffalo, who has created what he calls "life-structure" tables of ancient populations, extinction could have been caused by a "very, very small amount of change." He examined models of both modern man and the immediately preceding Neandertals, and found that a mere 2 percent increase in mortality could have led to rapid extinction in an archaic people, wiping them out in just thirty generations—or 1,000 years. The extinction of the archaics, according to Zubrow, could have been due to disease, competition for resources, or, yes, "actual warfare."

But Richard Leakey, who thought the DNA evidence presented "one of the most interesting questions before us," discounted a holocaust as "overdramatic." Prior to firearms, humans didn't have the capability of annihilating *any* species, he says. Rather, it seemed as though European archaics, which were best represented in their latest stage by Neandertals, simply faded into the background. "Replacement can come in several forms: competition for range, cross-breeding, absorption," he says. "And I think if it was *Homo sapiens* as opposed to a separate species, that you are going to have Neandertals and other populations mixing at some point in time. And I think you have lost it within rather than causing its extinction by the more dramatic and exciting concepts that Wolpoff and others are suggesting."

Instead, says Leakey, the factor causing one group to dominate over another may indeed have had to do, as Zubrow suggested, with disease. "Certainly the archeological evidence would suggest that there was remarkable homogeneity of the known population, which was presumably late *erectus*/early *sapiens* across most of Africa, a good part of Europe, and a good part of Asia and the Far East until about 300,000 or 200,000 years ago, and thereafter you seem to get new technologies, new ways of doing things," he says. "It's very difficult to interpret that, but it does suggest that something happened about then. Now maybe this is an indication within that sort of general time frame that you've got a demise of most of the population perhaps from some retrovirus like we're seeing with AIDS. We've seen it in other species: social carnivores with distemper, we're seeing it in the cheetah. So it could have been something that pretty much eliminated a large part of the population except a pocket of primitive people, and that pocket then had the resistance to whatever it was, and what you're picking up in the mitochondrial DNA is a reflection of an event of *that* kind."

But the idea of a killer horde rushing around the Mediterranean to eviscerate and for all anyone knew cannibalize archaic populations in Europe and Asia was too much for any sensible journalist to pass up, and there it was, mentioned in no less than a cover story in *Newsweek* on January 11, 1988. On the front of the magazine was an enthralling color portrait of a gorgeous, chocolate-colored couple in the Garden of Eden, while inside was a picture of a Rebecca Cann who looked rather sour at the chore of posing for a photograph (and perhaps not overjoyed at being on the front line), along with another of antagonist Milford Wolpoff looking serious at a table full of the skull casts he pulled out to prove his points. "Did the immigrants kill the natives?" asked *Newsweek.* "Possibly, but the conquests may have been peaceful."

Not according to Wolpoff. Not in the view of Harvard's Ernst Mayr, either. "I don't know if there is anywhere, even in modern times, where there was replacement of one group by another without some violence," Mayr points out. "Can you give me an example? Whether it's North America or South America, or Central America, or whites coming to Australia or even going to Africa, there's an awful lot of turmoil and bloodshed, and the more primitive they were, the more likely they were to fight for their existence."

Back toward the beginning of this century, fossils found at a rock shelter in the town of Krapina, Yugoslavia, were perceived as displaying a meld of modern and archaic features that led researchers to hypothesize that the bones were from a clash between Neandertals and modern men for control of the rock shelter. That was even better than a holocaust: a pitched battle. Wolpoff himself had written of the *erectus* skull wounds. "Many Neandertals have fractured skull bones, indications of violence," adds Alan Mann, a Berkeley graduate but a Wolpoff ally. Outcompetion didn't seem very reasonable to Mann, and it was his view that in order for the Berkeley theory to be correct, "a pretty scary conflict" had to have been perpetrated by *Homo sapiens sapiens* against more primitive *sapiens*. "They wiped everybody out, to a child. There was not a single genetic contribution of those archaic *sapiens* to the modern line. Cann, Stoneking, and Wilson say they could spot that and it simply wasn't there. If that is not true, then you can't really take their work seriously."

The scenario was impossible to resist: sleek, rather gracile humans from the south who had forsaken the browridges, whose arms were no longer heavy and stubby, who were no longer covered with matted hair, moving in throngs up the east coast of Africa across the badlands of Ethiopia and the Sudan, thereafter across wistful Egypt and into Israel and Iraq to do battle with thick and low-skulled archaic holdovers, nailing them with hafted stone points while the archaics could only toss back a rock and growl.

Certainly there was some new technology involved. Were the old, *erectus*-like *sapiens* eradicated with new blade tools? Or captured in traps?

Then the Africans massed east to annihilate the archaic *sapiens* of Narmada, India, and anyone in China who looked the least bit like Peking Man.

Bury my heart at wounded Zhoukoudian! Meanwhile a large splinter group of Eve's sons had surely headed north through Turkey and into central Europe or turned west to trample upon the descendants of archaics near Cava Pompi and Fontechevade, right through La Chaise, like swarming Indians, bludgeoning the Neandertals who had inherited the French terrain from other, older archaics, up through Europe spreading in every direction—north 34,000 years ago to settle Picken's Hole and soon Kent's Cavern until they had destroyed the shamblers from the Channel Islands to Pontnewydd in North Wales.

Wolpoff smelled blood, and it wasn't just from some Pleistocene "holocaust." Really, few paleoanthropologists believed it had happened. There may have been some battered archaic skulls, and cannibalized *erectus* brains, but they had probably done that to each other, perhaps in ritual. Nowhere was there evidence of warfare.

The blood Wolpoff smelled was really what he perceived to be Berkeley's increasing vulnerability. One of the first indications of that had come in their Cambridge paper. Cann and Stoneking seemed two-minded about admixture, not sure if it had factored in after all, and they had stepped back from the earlier pronouncement of 142,500 to 285,000 years ago for Eve, which seemed so authoritative and precise, and were now saying 50,000 to 500,000 years.

There was an order of magnitude in there. They were opening up a range of 450,000 years. It was nearly as fluctuant as the Mauer jaw or skull 1470. That seemed to Wolpoff like waffling. "It also seemed like a wild guess."

Maybe they were getting a little scared. There were all kinds of qualification coming down. They wanted to have their cake, he thought, and eat it too. On the one hand they would say the migration from Africa could have been as recently as 23,000 years ago, on the other hand Cann was writing that Eve's offspring may have gotten to Asia 140,000 years ago, since that's when divergences in DNA indicating a partitioning of lineages started to appear.

Yes, thought Wolpoff, they seemed to have spoken too soon. They didn't have a good grasp of the fossils. Their hypothesis lacked cohesion. He wanted to know if natural selection had acted upon the molecule to skewer its traceability, if Eve was just random survival of a mitochondrial lineage, and if the mitochondrial DNA was really as neutral in evolution as he perceived it as having to be if their assumptions were to hold up. Their very analytical model remained unproven: a 2 to 4 percent rate of mutational difference per million years. Even Sarich, in their very own lab, thought their calibration of time was off base. Way off base. As far as Sarich was concerned (see Notes), the first evidence of people coming out of Africa was 700,000 to 800,000 years ago, the old *erectus*, heading for China, while Cann, et al., at least in the *Nature* story, were saying African progenitors had not gotten to Asia until half a million years later.

Stoneking was sedate as usual, unemotional about the debate, and Wolpoff or no, he didn't seem very frightened to me. He was matter-of-fact about Wolpoff, as he was matter-of-fact about everything. "I think if we would have shown that the common ancestor for all humans was in Africa and lived 600,000 to a million years ago, everyone would be very happy with that, because that would correspond to *Homo erectus* or something like that," he says. "Using our present calibrations, which admittedly are crude but are the best we can come up with at this time, we just can't push the date back that far."

They were still sticking to the basic date of 200,000 years, and the debate was now an international one. In England, paleontologist Chris Stringer was using the data to support his concepts of European settlement, and so was Günter Bräuer in Hamburg. Stringer was Wolpoff's arch-rival. Then you had some Japanese geneticists finding a different rate of divergence while working with the same mitochondrial DNA.

Meanwhile Doug Wallace, the geneticist from Stanford, had found, contrary to Cann, et al., that Africans, Asians, and Europeans all had characteristic mitochondrial DNAs. Berkeley had not found distinctive DNAs associated with races; the branches on their tree held that assortment of ethnic types.

Said Wallace: "Specifically, one mitochondrial DNA lineage seems to have survived from the primordial human population. As the female descendants migrated to the various continents, they became isolated. New radiating lineages then developed by the progressive accumulation of mutations along the maternal lineage. Since two mitochondrial DNAs appear to be central to the tree and are found on all continents, they may well be representative of the mitochondrial DNAs from our original female ancestors. This theory is supported by the fact that the DNA sequence of diagnostic restriction site of one of these mitochondrial DNAs is identical to that of all great apes but different from all other human mitochondrial DNAs."

Although the publicity had started out slowly, newspapers couldn't resist a story about the discovery of Eve. The DNA hypothesis was treated seriously by important journals like *Science*. Paleoanthropologists had introduced the term "mitochondrial DNA" into their already overburdened lexicon. It was quickly a new orthodoxy in certain circles of anthropology, and lengthy review articles on the subject were being scheduled for weighty reference publications such as the *Annual Review of Anthropology.*

If it was a ripple at first, it had turned into something not unlike a tidal wave. The cover of *Newsweek!* By 1989, a random mention of the mitochondrial data to any of the nation's major paleoanthropologists, archeologists, taxonomists, geologists, or paleontologists would evoke a galvanized response. It was about the most heated controversy around, and along with recent fossil material that had surfaced lately, it had pushed debate about modern man back into the forefront, stealing thunder from the usually more intriguing man-apes. The idea of an African origin 200,000 years ago became a subconscious reference point, just as paleoanthropologists had come to nearly subconsciously accept the idea of a splitting of man from the apes 5 million years ago, and it was a topic everyone had an opinion about. It may not have been boiling over the kettle like the debate between Leakey and Johanson, or utterly in the alley like the rancor between Sarich and Sibley, but it was getting there, and the kettle was a larger one.

It was the big time. There was almost the sense that Becky Cann was going to be the Mary Leakey of genes. Wilson was the genetic version of old Louis, but far more secretive and unengaging, less colorful, at least for now, but equally sure of himself. Who needed fossils when you had mitochondrial DNA? A new day was dawning for anthropology, and the sun that would shine would illuminate man's tortuously convoluted ancestry.

That is not what Cann was feeling or certainly not anything she overtly thought. There really wasn't the publicity hound in her. Cann was already tired of talk shows because they cut down on her time to "do science." It was just that she had worked years on this project, it was part of her very dissertation, laboring through many a night with what Wes Brown had started, egged on the whole way through by Allan Wilson, who was always looking for new horizons to conquer, new theories to announce, and now somebody had to explain all this. After all, major funding had been in United States tax dollars.

Somebody also had to defend the hypothesis, and Cann was showing the strain of the trashing by Wolpoff. He had accomplished part of what he'd set out to: he was under their skin. When she saw an article written for a journal called *Mosaic* that had given

Wolpoff substantial play, she responded with what Alan Mann describes as an "offensive" letter and what the writer himself, Arthur Fisher, who works for *Popular Science*, describes as unlike any letter he'd received in twenty years of science reporting. "She seemed quite overwrought," says Fisher. "Her comments were really very extreme. To just give you one example, I had a sentence about Wolpoff in which I said he was a noted expert who had handled more hominid fossils than any other anthropologist, and she inserted 'handled and *broken.*' That gives you one small example. She certainly cast aspersions on my ability as a science writer. Basically she was upset because her theories were being held in question by other people. I've never seen a reaction like that."

"She lost control," says Mann. "And I think you can't do that."

But there was so much misunderstood by reporters, the Berkeley group felt, and the press was a pain in the ass. Even when there weren't inaccuracies, Stoneking felt uncomfortable with things like the cover story in *Newsweek,* which he criticized for concluding that the genetic search might soon turn up Adam as well.

Yet at the same time they were criticizing *Newsweek*—making jokes about its cover, Stoneking disturbed about the mention of an Adam—Cann herself had proposed that "a comparison of mutations in the male sex chromosome could generate a genealogy leading back to Adam, and ultimately could determine whether he lived at the same time and in the same place as Eve."

As for the paleontologists and paleoanthropologists—who were they to talk? That's what Cann wanted to know. Here they asked everyone to accept the notion that they could piece human ancestry into neat categories when what they were really doing was basing their ideas on assumptions that entered into their evolutionary reconstructions from isolated teeth, shattered skullcaps, and woefully fragmented jaws. "Many paleontologists fear that if they expose the legitimate scientific limits of the certainty of their theories, fundamentalists and creation 'scientists' may misrepresent these data to dispute the fact that evolution occurred," she charged.

Wilson still could not be bothered. His style was to hide

away collecting data, package it for various journals, make select appearances to get a new idea in motion, which put the publicity on his own terms, then hide away again as the data and hypothesis detonated, professing to care nothing about publicity—acting in fact as if he disdained it, which he may well have, or at least the process of obtaining it—but allowing his work to be phrased in a way that guaranteed maximal exposure. It may have been an ache in the butt, but all those news stories were probably not bad for funding.

Not that that was necessarily Wilson's motive. He was a serious scientist, an unorthodox scientist perhaps but a respected scholar, winner of a Guggenheim, a native New Zealander who'd taken the world of molecular evolution by storm, first by upsetting the notions of the ape-human split with his protein workups, now moving swiftly up the ladder to the area of greatest interest, man's point and time and circumstance of modern origin. All about Eve. It was history's great breakpoint. All they needed to remain in the ball game was a constant mutation rate of 2 to 4 percent and for the mitochondrial DNA from their given time period to have really been representative of a common ancestor who symbolized the threshold of modern man. They also needed someone to explain how Eve's kind had taken over the world without a holocaust. Wilson seemed relaxed about it all.

But what about Becky Cann? She could hold her own in just about any intellectual setting, she had a hard edge that cut through niceties, and she was pretty feisty herself, but Wolpoff was coming with a wrecking ball. He was even attacking Stephen Gould, who liked the way DNA fit into his theory of punctuated evolution (since it seemed to indicate "a speciation event, a branching event in one small part of the range of *Homo erectus*") and who thought the concept of a mitochondrial Eve and a recent origin for all of mankind was a nice and gentling hypothesis because it spoke of "a kind of biological brotherhood that's much more profound than we ever realized."

"Gould says that if nothing else, this shows the close relationship of all of mankind," retorted Wolpoff. "And I think to myself: here's a theory that a group of people went out and killed everyone off, and Gould is telling me this is a definition of brother-hood!"

When Cann jumped back in, it was to explain that anthropologists were threatened by the new technology. Their beloved fossils were at risk. They couldn't understand it all. Their grasp of biology was "very poor" and what they knew about genetics they had learned in the 1950s.

"Mitochondria don't pay attention to the rules for nuclear genes and require a different kind of thinking," she thought. "When anthropologists teach biological anthropology they talk about Mendelian genetics. That's not what this stuff is! They don't have a secure frame of reference for it. It's foreign, it's different, it's 'quasi-credible.' We've got the limbic system going here, not the cranium that's responding. Many zoologists, mammologists, and ichthyologists are using mitochondrial DNA to explore ramifications of population movements and dispersements, and they don't have any trouble with mitochondrial DNA. It's when you talk about humans and the fossil record and *ourselves* that all these issues become controversial."

And if their "beloved" fossils lost some of their meaning through this confusing new way of looking at genes, the paleoanthropologists might in turn lose some of their funding. That was at the crux of it. In trying to raise funds, people tended to become "more outrageous in their claims, and ego-ridden," she claimed.

There *was* a lot of money involved, Mann pointed out, there was a ton of prestige. Everyone had a theory on man's origins. The mailman had a theory of man's origins. "Human evolution is a big deal these days," noted Mann. "Leakey's world known, Johanson is like a movie star, women moon at him and ask for his autograph. Lecture circuit. National Science Foundation: big bucks. Everything is debatable, especially where money is involved. Sometimes people deliberately manipulate the data to suit what they're saying."

Mary Leakey partly blamed the media for the controversy by hyping the concept of an Eve. "I thought, how silly. It's not the way to go about it nor to regard it. I think people have gotten overenthusiastic, as they often are when things are new. And they are probably claiming more than they can at the moment. But once it stabilizes I think it can be *extremely* significant. It's now fossils versus DNA and they start fighting immediately. But I see no reason why they are incompatible."

The other elderly statesman of African paleoanthropology, Phillip Tobias, was less serene. No Asian input? "I find that aspect quite inconceivable. We have so many forms from China, from Java, from Australia, from Africa, and from southern Europe which look intermediate between *erectus* and modern *sapiens* that I can only think that there was a kind of worldwide gradual transition with key mutations arising in key spots, but I can't really find any biological feasibility in the suggestion of one small single group arising in Africa and pushing out—totally superseding—all of mankind in Europe, all of mankind in Asia, all of mankind in Australia and everywhere else."

There was no way, thought Tobias, whose denunciation counted far more than most, that mankind's history was going to be contained in a tiny portion of genetic material. "Key mutations may have arisen in key areas and then gradually diffused through the local populations and adjacent populations in ever-widening circles, but to create a story of the whole of mankind just on the basis of a single DNA nuance I find just as unacceptable as the earlier investigators who tried to reconstruct the whole story of humanity on ABO groups—the red blood cell," Tobias says. "They used one criterion and built up flamboyant histories of human migrations around the globe with no relationship to what they looked like in other parts of their body."

It was really interesting that geneticists could scoff at the fossil record, thought some, when fossil morphology, by many measures, was a task more complex and more taxing than restriction sites or gene sequences. It was also fossils that provided points of calibration.

"Sadly, the fossil evidence doesn't help refute or support the DNA work" was what Richard Leakey had to say. "One would hope that the fossil record and the DNA record will ultimately tell the same story. It would be extremely sad if they didn't!"

Gould kept his distance, fearing the "emotion of it all," while Leakey was trying to defuse any polarization between geneticists and paleoanthropologists. He shouldn't have worried. It wasn't going to be a simple war between two disciplines: paleoanthropologists were already lining up on opposite sides of the DNA debate and geneticists were also arguing among themselves. For those disinclined to the Berkeley claims, there was the distant image of know-it-all geneticists at a school where old hippies still wandered by with flutes and where that Xerox place (2½ cents a page) was the Krishna Copy Center.

"Perhaps their interpretation of some of the genetic material is made more secure than it actually is," complained Fred H. Smith of the University of Tennessee, who was "upset that there is the impression out there that geneticists have all the answers. I think they regard their evidence as somewhat more precise and therefore perhaps more meaningful than some of the gross structures we're looking at. But the fact of the matter is that the only truly direct evidence that we have to connect together living groups is what is left over from the past, and that's the fossil record."

Wolpoff's rival Chris Stringer, perhaps Britain's leading authority on the rise of modern man, had immediately embraced the DNA work and had formulated his own new theory to countermand what Wolpoff and allies such as Smith said, marshaling his version of the fossil record to support the out-of-Africa hypothesis, and agreeing with Cann that the debate came down to paleoanthropologists trying to keep their fossils from "being tossed in the bin. That is what the mitochondrial DNA at its clearest and simplest is saying to them. It's saying, 'Your fossils are not ancestors of modern men. What you have done for the last ten, twenty years was a complete waste of time.'"

17

MOST CONTENTIOUS OF ALL was the fate of the Neandertal. By granting the impression that it too was not a direct ancestor, the DNA hypothesis managed to offend not only paleoanthropologists (whose careers had been spent studying this archaic from Europe), but more importantly, the theory offended the public at large. To the average person, "Neandertal" was synonymous with the term "caveman." For the vast majority of the world, it was the definition of prehistoric man.

There it was in outdated dioramas: *Australopithecus* to *habilis* to *erectus* to Neandertals.

Yet Neandertal people, said Cann, were not thought to have been ancestors of modern humans and one reason was because they were simply not as old as the oldest *Homo sapiens sapiens* from Africa. According to Berkeley, modern humans (as opposed to the archaic *sapiens* ancestor) probably developed in Africa 100,000 to 140,000 years ago and all present-day humans were descendants of that population. Neandertals, on the other hand, had been confined mainly to Europe, originating at the most 125,000 years ago. Even if Neandertals were as old as Eve's clan, Cann had said, "it would be difficult to explain how they kept essentially the same form for 60,000 years, then, overnight, evolved into a new subspecies 40,000 years ago, when modern humans first appeared in Europe."

One thing seemed certain: by 30,000 years ago, the Neandertals had quite suddenly and mysteriously disappeared. Or at least most of them had disappeared. There were pockets that may have survived for a few thousand more years. But if so, they were not in significant numbers, according to most paleoanthropologists. Having first arisen from older members of the genus *Homo* more than 100,000 years ago—apparently spawned by those many archaics from Biache to Swanscombe who had shown pre-Neandertal traits—the Neandertals had managed to survive in Europe as well as the Middle East for about 2,850 generations, making it through glacial weather, fending off the woolly mammoth, and then: curtains. In their place were the Cro-Magnon men who decorated caves in the south of France and who, unlike the Neandertals, were anatomically modern.

If Berkeley was correct, the Cro-Magnon specimens (*Homo sapiens sapiens*), which dated in Europe to a time period right around the demise of Neandertals (*Homo sapiens neanderthalensis*), probably represented Eve's descendants once they had implanted themselves in Europe. "Cro-Magnon" had become a broad term for the first modern Europeans, and although the specimens that were actually found at the first Cro-Magnon site, near the village of Les Eyzies in southern France, were perhaps 25,000 years or younger, other examples of modern humanity in Europe went further back in time than that—as Cann said, to about 40,000 B.C. Cro-Magnon was, in metaphor, one of Eve's proliferating grandsons.

According to Cann's scenario, Eve's children had probably outcompeted the Neandertals, and the key impression, again, was that the two groups—moderns from Africa, Neandertals from Europe—did not interbreed and were not directly related. You could also invoke the scenario of violence: if the two groups did not get along well enough to interbreed, then perhaps they did indeed fight with each other.

But again there were no signs of such violence, and it was much more likely a case of outcompetition, the "conquering" perhaps taking 5,000 years to accomplish even if it seemed in retrospect so very "sudden." The idea that Neandertals had not served as our direct ancestors was anything but new, debated now for eighty years. But that they may not have had any genetic input whatsoever into modern men still struck one of the world's foremost authorities on the Neandertal, Erik Trinkaus, as "the extreme point of view" and one that, as such, was almost surely wrong.

"The idea that we had a very localized group of modern humans which emerged in Africa and then were the sole ancestors of all modern humanity, I think that's simply unsupportable by the evidence," complained Trinkaus, a University of New Mexico anthropologist who took the debate a step further, noting that if certain of the assumptions Berkeley had used were altered—for example those involved in the method of rooting their evolutionary tree—the time depth for Eve's existence could be doubled or tripled. And if the time of Eve's existence was doubled or tripled, bringing her back, at one extreme, to 840,000 years ago, "then all [Berkeley] would be documenting is the spread of *Homo erectus* out of Africa. And if that is indeed the case, then the mitochondrial DNA data, while very interesting, has no relevance at all as to the origin of modern humans."

Nearly immediately after discovery of those thick thigh bones and the barbaric skullcap at Feldhofer Grotto in 1856, the Neandertals had been relegated to a taxon separate from modern humans, and by the turn of the century there generally were two possibilities recognized for their place in evolution: that Neandertals were in between Java Man (*erectus*) and modern humans—a missing link—or that Neandertals were more of a side branch of the human lineage, an irrelevant offshoot. Around the time of World War I, a strong intellectual feeling had grown in western Europe that Neandertals were too specialized to be in man's ancestry at all. This was known as the "pre-*sapiens* theory," and the idea was that the lineage leading to modern humans went way back in time and was basically unrelated to development of Neandertals. It was a theory that greatly benefited from the Piltdown skull, which seemed to show modern traits long ago, preceding Neandertals. Of course, Piltdown turned out to be a fake, and "pre-*sapiens*" theorists had to look elsewhere.

In opposition to the "pre-*sapiens*" theory was a "unilinear" point of view, which held that mankind had evolved in a direct line from *erectus*, archaics, and Neandertals. This school of thought is most similar to what is now known as multi-regional continuity, and its early adherent, Franz Weidenreich, saw a Neandertal phase in human development although he also figured that the actual transition into anatomically modern *sapiens* had occurred in western Asia.

When there were attempts to claim the archaic fossils from Vértesszöllös, Hungary, were *sapiens* that went way into the past, demonstrating an old lineage for modern man separate from that of Neandertals, as the "pre-*sapiens*" proponents had espoused, a young Milford Wolpoff had stepped into the debate, arguing, in 1971, that Vértesszöllös bore resemblances to *erectus*, which meant it was not the "pre-*sapiens*" lineage that the opposing school was so ardently searching for. A similarly strong proponent of continuity, C. Loring Brace, also of Michigan, argued that the robust Neandertal face turned more gracile and thus more modern as more specialized tools replaced tasks handled previously by the Neandertal's anterior teeth. It was an explanation, in other words, for how this time-honored form, wide between heavily browed eyes, beaked of face, thick of jaw, could have turned modern over a relatively short period of time.

One thing was no longer a point of hot contention, and that was the idea that both Neandertals and modern *sapiens* rose from the *erectus*-like archaic form. This we know for a fact. Indeed, there was so little difference between the Neandertals and certain of the older European archaics discussed in Chapter 13 that occasionally there had been confusion over which of the two subspecies they belonged to. Some paleoanthropologists even used the term "Pre-Neandertals" for the archaics. Specimens such as Swanscombe and Biache belonged to the archaic subspecies, for example, but showed distinct tendencies toward the Neandertal form. At the same time, certain of the older archaics in Spain, Germany, and Italy, in addition to England and France, had shown traits at least superficially forerunning *Homo sapiens sapiens* while also harboring features that would one day manifest themselves in the form of Neandertals.

Although the brows of Neandertals were slightly lessened, and their brains were larger, they still bore a great number of resemblances to the archaics and the classic *erectus* that had come before the archaics, their bodies big-boned, heavily muscled, and bulkily built. The stout, strong physique may have been testimony to arduous days of foraging and hunting. Their skulls, though holding brains bigger than most archaics, were still lower and flatter than a current-day European's, but much thinner than an older archaic type of skull. Like the archaics, the forehead remained in retreat. The chin was still fairly weak but getting stronger. The big difference was at the back of the head, where the Neandertal skull bulged into what is known as a "chignon," or that occipital bun. The face of the Neandertal also jutted out, especially at the nose, which was nearly a muzzle. Unlike the *Homo sapiens sapiens* who were also coming into existence and would succeed them in Europe, the Neandertals, besides bone thickness, were also long of jaw.

It was such robust characteristics, especially of the brow, that had caused the discoverers of the first fully analyzed Neandertal near Düsseldorf to assume that it was a barbarian or perhaps an ape-man. Or a pathological Cossack chasing Napoleon! This was before *erectus* had been discovered by Dubois in Java, and so no one yet knew what an ape-man really was. In fact, there were morphologists who subsequently described the postcranial differences between Neandertals and modern humans as trivial. They had the same neck lengths as modern humans, vertebral curvatures well within the modern range, and intervertebral articulations identical to those of *Homo sapiens sapiens*.

So it looked like the archaics of Europe were ancestral to both Neandertals and the modern men who followed them in western Europe. But according to Berkeley, that seemingly logical notion was also wrong. The Wilson lab had basically rendered both Neandertals and the archaics from Europe as irrelevant, once and for all knocking down the idea of a unilinear and continuous transition from archaics to Neandertals to present-day Europeans. Most startling was their elimination of European archaics who could as easily have given rise to modern physiology as the archaics in Africa.

The reason it was so startling was because there seemed to be archaics in Europe that had been heading in the direction of modern man, and certain specimens of Neandertals that combined archaic traits with modern ones—which made it seem like Nean-dertals were more than just a dead end. As in Asia, for Berkeley to be right there could be virtually no genetic continuity between old archaic Europeans dating to the period before the African "invasion" and modern Europeans. And for Berkeley to be right, Neandertals shouldn't have too much in common with modern man because Neandertals, said Berkeley, had left behind no genes—just as Peking Man supposedly had left behind no genes.

Yet such scholars as Ralph Holloway of Columbia University viewed Neandertals as different from modern people only in a way that was analogous to the differences between Australian Aborigines and Eskimos: in other words, an ethnic variation of Homo sapiens. And treating the Neandertals as hapless brutes who had nothing to do with modern man, said Wolpoff, taking this a step further, was nothing less than racist. Besides, he said, there were Neandertals who bore resemblances to us.

After the discovery near Düsseldorf, Neandertal remains also had been found at a cave in Belgium and near the village of La Chapelle-aux-Saints in France. Indeed, so many Neandertals were found in the Dordogne and other parts of southern and central France, including a famous rock shelter at La Ferrassie, that it appeared this part of Europe had been nothing less than an ancient population center.

One of the French caves, La Mousterier, not only yielded the skeleton of a Neandertal youth but had also provided archeologists with important stone implements. The culture of Neandertals would thus become known as "Mousterian," just as the one associated with erectus was "Acheulian." And impressive these tools were. While the archaics immediately preceding Neandertals had been adept enough to advance toolmaking to the point where stone flakes were becoming saw-toothed and regularly grooved, and the hand axes more intricately chipped, the Neandertals improved upon the methods with a style whereby a nodule of stone was trimmed around the edges to make a disk-shaped core and then pounded into sharp-edged flakes, with a variety of new and more specialized forms. There may have been sixty different types of stone tools. Neandertals used flint points, picks, flakes, leather punches, and scrapers, which were canted or convergent or double-edged. Some of them looked nearly lathed, requiring a lot of precise hammering. Retouched flakes became the preeminent tool, and there were very sharp flint knives.

Then, about 35,000 years ago, according to archeologist Desmond Clark of Berkeley, there was an abrupt shift in technology on the European continent that may very well have heralded a technology imported from elsewhere—such as Africa. Not everyone thought it was so "abrupt," and it may not have offered revolutionary change, but if true, it was a spellbinding fact, because 35,000 years ago was also the time Neandertals "disappeared" from Europe and anatomically modern humans emerged in the form of Cro-Magnons.

This "new" technology was notable for the production of blades, and forerunners of such blades have been found in South Africa dating to approximately 75,000 years ago—many millennia before a version of this finds it way into Europe, according to Clark. Other precursors to the technology which so "suddenly" appeared in Europe around the time of the Neandertal disappearance were to be found in Sri Lanka and then in southern Israel, perhaps dating to 40,000 years ago, just before a similar technology entered Europe. Did modern genes accompany new tools from Africa, wending through the Middle East? And did those genes represent Eve's?

It would have fit in the sense that Cro-Magnons had longer arms than Neandertals and long-armed people are likely to have come from a warm environment like Africa. Short arms were supposed to conserve heat in the cold, lessening loss of warmth from the limbs, and the Cro-Magnon type was able to compensate for this, according to some scholars, by making warmer clothes (sewing was coming into style then) and building better hearths. Moreover, the increased stature may have allowed modern men better locomotor efficiency, according to Trinkaus and Fred Smith, suggesting that "the early anatomically modern humans were better adapted to covering large distances and hence able to monitor more effectively and exploit more efficiently the available resources with less physical durability and stress."

In other words, who needed a barreled chest, a big torso anymore? The moderns had *style*. Style and grace. It was once thought that perhaps females were pregnant for a shorter period of time (as opposed to the eleven or twelve months of gestation for Neandertals), and this, speculated Trinkaus and Smith, would have allowed a baby's brain to develop faster, since it was exposed sooner to greater stimuli, which led to a higher and more rounded head, as well as allowing a woman to get pregnant again sooner, which may have led, in *its* turn, to faster population growth. Trinkaus had retracted the interpretation of gestation after reviewing new fossil evidence, and Smith was not certain any longer that a modern body would have been better equipped for traveling the rugged terrain than a Neandertal's; but, anyway, there was something about a gracile body that lent to visualization of more athletic movement (the Neandertal was more wrestler than marathon man) and a lithe physique better fit the lithe, modern tools of the "Aurignacian" or Upper Paleolithic industry, which soon would include hafted tools (if not already so) and implements that were no longer so clumsy, offering more mechanical leverage and requiring a precision grip. As for the face and skull, it no longer had to anchor the sort of mouth that had been used as a vise by the Neandertals.

"What you can see of course with the appearance of the Upper Paleolithic is something that is very different from the technology associated with the Neandertals, which is essentially a Middle Paleolithic technology based on flakes and certain types of prepared cores, whereas the Upper Paleolithic was based upon blade technology and the whole range of blade tools and of course a whole lot of bone/antler tools as well—indeed, ivory," says Desmond Clark. "A whole new technology arises between 40,000 and 30,000 years ago. Extremely interesting. In the Negev [southern Israel] you have true Upper Paleolithic technology—you can see it emerging out of the Middle Paleolithic technology around about 45,000 or more years ago—just before Europe."

If the current collection of Neandertal fossils, representing more than 200 individuals, has been correctly interpreted, and if dates suggesting that Neandertals may have persisted in some European spots until much more recently are as wrong as they are suspected of being (see Notes), the Neandertals disappeared during the last period of Ice Age coldness. Although it seems unlikely that the last stages of such coldness had dramatically affected the Nean-dertals, since they were so used to such a clime, there were those who thought that perhaps the climate had pushed many of the archaic types down from Europe and into warmer territories, which, says Clark, may have created "a certain amount of conflict heightened by, say, a glacial advance and drier conditions on the Sahara." The less available good land, the higher the chance of violent interactions.

Although most Neandertals were associated with the Mousterian culture of that archeological era known as Middle Paleolithic (which was still part of the Old Stone Age), there have been indications—for example, at Saint-Césaire in France—that some Neandertals were also associated with the next era of tool-making in the Upper Paleolithic, sharing certain of the technolo-gies with the anatomically modern humans whose stylistic and functional changes in stone and bone equipment, as well as in-creasingly complex techniques of food processing, indicated, ac-cording to Desmond Clark, a more efficient economic and social organization. In other words, some Neandertals had developed new technologies. One nearly feels obliged to visualize Neandertals watching the new incoming humans at work and learning to imi-tate the imported technology. "People are suggesting that this may be the attempt by Neandertals to produce an Upper Paleolithic technology, to adapt to an Upper Paleolithic technology," Clark says.

But the innovations of the Upper Paleolithic were not all that superior to what Neandertals were used to, argue others, and the cultural reflections were quite at odds with the lampoons of a knock-kneed ruffian—the classic caveman—dragging his girlfriend by the hair back to the cave for a nefarious thing or two. For one, most Neandertals probably didn't live in caves but rather in the open, almost surely in skin-covered abodes superior to the huts at Terra Amata. The caves were simply where bones tended to be better preserved. And a lot of the bones, coming from reindeer and other fauna, suggest that Neandertals were fairly good hunt-ers. Along with greater dexterity—which was indeed impressive considering the Neandertals' stubby fingers and hands—these an-cient Europeans who were also to be found in such places as Israel and Iraq had developed rituals and a spiritual awareness far more definite than the dubious hints of ocher decoration in the archaics who had come before. The Neandertals buried their dead and apparently also presented grave offerings and decor with the corpses, including mammoth tusks. They also took care, it seems, of the elderly and infirm: one old man was missing a forearm. Uncared for, he would have succumbed to the elements. Goat horns have been recovered from a boy's grave in central Asia, and flowers (identified from their pollen) in a burial at Shanidar Cave in Iraq, note Trinkaus and Howells.

Though it was the subsequent Cro-Magnon Man who developed the first truly noteworthy examples of art, Neandertals have provided hints of personal adornment and may have etched bones with incipient and not altogether decipherable designs. Strong of tendon, short of limb, the Neandertals, according to Trinkaus and Howells, probably formed hunting bands linked loosely into tribal groupings. "To judge by the wide distribution and homogeneity of Neandertal remains," noted the two scholars, "the Neandertals formed a distinct and major human population that was not a particularly sparse one."

And so the question again rears its head: How did they so suddenly disappear? And did they really disappear, or did they evolve into us?

The Eve hypothesis seemed to lay the matter to rest: Neandertals were not ancestral after all. In fact the very way they vanished suggested, to some of those who subscribed to the DNA model, that it was indeed Eve's children who one way or another shoved Neandertals off the face of the earth.

The Neandertals had developed over the course of about fifty millennia, had remained relatively stable for another forty millennia or so (according to many anthropologists), and then, over the course of a mere 5,000 to 10,000 years, had turned into—or given way to—anatomically modern humans like Cro-Magnon. Was not an invasion from Africa a likely explanation, as Cann said, to the conundrum of such a seemingly rapid turnover? And what about the tools? According to Trinkaus and Howells, between 35,000 and 40,000 years ago there was a marked increase in the complexity of their sociocultural system, with art becoming a regular feature at archeological sites. Didn't that too imply an outside influence that was modern?

It may have, but to many paleoanthropologists, including some major ones, Neandertals were no genetic dead end. First of all, argued people like Wolpoff, the Neandertals were not a species that had remained stagnant for most of their existence, as his opponents claimed, but were rather a species showing certain signs of modern evolution. After studying Neandertal specimens from Yugoslavia, he saw a clear progression from more archaic Neandertal physiology at that place called Krapina to more modern form at Vindija, also in Yugoslavia. The Vindija Neandertals, wrote Wolpoff, "although still within the Neandertal range, give the impression of being nearer to the early modern European condition than are most other European Neandertals. The Vindija specimens exhibit reduced midfacial prognathism, reduced facial size, thinner and less projecting supraorbital tori, a tendency toward smaller anterior teeth (especially in breadths) . . . and they seem to lack occipital bunning."

At the time Neandertals were supposed to have "disappeared," there were a number of populations, best represented by a frontal bone from Hahnöfersand, Germany, that seemed to be transitional specimens bridging the gap between Neandertals and modern humans. Hahnöfersand was unusually thick of bone, even by Neandertal standards, and had a sinus just like the famous cavemen but had a modern browridge. Clearly there was the possibility of hybridization between Neandertals and moderns on the order of 35,700 to 36,900 years ago, the date of this sample. It seemed to tie together Neandertaloid traits found too in France and down in the Middle East. "Summing up the result," wrote Günter Bräuer of the University of Hamburg in the *Journal of Human Evolution*, "one can conclude that Hahnöfersand has modern and Neandertaloid affinities. It is certainly difficult to state which affinities are dominant in this frontal."

Such features implied that some Neandertals were transforming into modern Europeans—contrary to the Berkeley theory—or were at least interbreeding with them. Characteristics of transition also were suggested, though less securely, in a specimen from Sala in Czechoslovakia and at that Belgium Neandertal site discovered after the Neander Valley find and known as "Spy." At Shanidar, where Neandertals may have been killed by a rock fall, one of the specimens demonstrated modern parietal proportions, but this may have been due to artificial deformation. At Mladeč in the Moravian karst north of Vienna, more importantly, there were crania exhibiting features of the Neandertals but there were also slender body bones like those of anatomically modern men. The Mladeč material, only some of which survived the 1945 Allied bombing at Mikulov Castle, where they were stored, may be, in their mosaic features, one of the most dramatic refutations to theories of complete replacement—of an utter African takeover— that Europe has to offer.

There were also indications of archaics and Neandertals slowly but surely turning modern in the Middle East. In fact, the most dramatic transition—if transition it was—seemed to be taking place in southwestern Asia, and transitional humans from outside of Africa—from anywhere but the sub-Sahara—did not seem to be what Berkeley had in mind.

Even Cro-Magnon Man, the very type specimen for the first European moderns, had a face positioned unusually far from the middle portions of the vault—resembling Neandertals, according to Wolpoff—and one specimen (Number One) had what Wolpoff described as a Neandertal nose. Trinkaus didn't see it that way but agreed that another specimen, Number Three, had a very bulging occipital.

This was extinction with no interbreeding?

Central Europe seemed like a hot spot of "evolving" Neandertals, and could date the Aurignacian culture back to 40,000 years ago, suggesting it was one of the first places settled by a somewhat different people moving up from the south. The gene flow causing a modern transformation may well have been coming from the Middle East and into Greece, where, on Crete, a partial postcranium possessing modern characteristics but retaining certain archaic traits has been dated at more than 50,000 years. "On the basis of the fossil evidence and the earlier presence of modern humans in Eastern and Central Europe," notes Bräuer of West Germany, "it appears more likely that early modern groups slowly expanded into Western Europe via Central Europe. They certainly coexisted with the Neandertals for perhaps some thousands of years (probably often in different territories), and also mixed with them to some degree."

While Bräuer and Smith focused attentions on the middle part of Europe, C. Loring Brace concentrated on northwestern European traits that he felt showed—contrary to the DNA—genetic input by Neandertals. Brace told me he was about to publish a study in which he graphed the proportions of Neandertal and modern human skulls—length to height and width to height. What he found, says Brace, was that precisely the same features that typified classic Neandertal form are the features that typify the differences between northwestern Europeans of today and other populations. The northwestern Europeans came out of this study most dissimilar to Africans, Australian Aborigines, Asians, and American Indians.

"So that it's quite clear that the modern Europeans are closer to classic Neandertal in the non-adaptive portion of cranial form than they are to any living human population," claims Brace. "It's absolutely open and shut. We've digitized the contour and extracted the ratios and angles, and we come to the same conclusion. The shape of the rear end of the European cranium indicates that the most likely ancestor for a modern European would have to be something that looks like a classic Neandertal."

Even before publication of Brace's paper there were questions about the methodology of measurements he took, and so this had to be put on hold for the moment; but in addition to that and the prominent nose that Wolpoff has mentioned as another surviving Neandertal feature in Europe, there were traits such as that bulging at the back of the head which Wolpoff's former student, Fred Smith, says lingered in early modern humans from central Europe, proving that Neandertals could not have been completely replaced. "I think in central Europe we can demonstrate that we have late Neandertals that are to a very great extent good intermediates between earlier, more primitive Neandertals and early modern central Europeans," asserts Smith. "And with the earliest modern central Europeans we have a number of characteristics that are most reasonably explained as some influence from Neandertals—not from something external into this area. In weaker moments when no one is around to browbeat me, I may still say that the Neandertal component is the *major* component of modern European development."

Smith described the evidence for actual large-scale movement of Homo *sapiens sapiens* groups into Europe as "nonexistent," explaining that early Homo *sapiens sapiens* specimens in Europe showed no distinctly African features. "For example, they should have *no* occipital bunning and *no* indication of the characteristic European early modern human browridge form, because these features are not found in the Near Eastern or African early modern human samples."

While Erik Trinkaus had no problems with the idea of a population fanning out from Africa, the notion that such a localized group of Africans would emerge as man's sole ancestors was another matter entirely. He insisted upon admixture. "Early modern humans evolved in Africa some time prior to 50,000 years ago, and there was a gradual spread—not a migration—population by population, valley by valley, into North Africa by 40,000 to 50,000 years ago, and around that time they became established in the Near East," Trinkaus believes. "And then gradually, around 35,-000 years ago, they were into central Europe and by 30,000 years ago into western Europe, admixing with local Neandertal populations to a variable extent. In some regions this admixture might have been trivial, in other areas it might have been quite substantial, approaching fifty-fifty, and this admixture was taking place in the context of some fairly major cultural changes. It was the Middle to Upper Paleolithic transformation. As this was going on those biological characteristics that were advantageous became dominant ones, and this is what we recognize as early modern humans."

Meanwhile, according to Trinkaus, the less advantageous traits—many of which were carried by archaics and Neandertals—died out or were eliminated by natural selection, and what we ended up with was early modern humans who had some characteristics inherited from the Neandertals but whose main traits were the result of a gene inflow that carried important advantages. It was like those pebbles Leakey had envisioned as being tossed into a pond. Among the advantages of early modern humans, thought

Trinkaus, was the ability to better process information about the environment, knowing how to get food, and knowing where the wolves and cave bears were.

"All the work that I have done suggests that the early modern humans were more efficient in exploiting these resources than were the Neandertals," adds Trinkaus. "It is not known exactly how they did this, but the evidence suggests that the anatomically modern humans had an energetically more efficient system of doing this."

But nothing so vastly superior to Neandertals, argue many scholars, that Neandertals, upon encountering more modern men, would have immediately bitten the dust. Indeed, archeologist Lewis Binford says that "one doesn't have to migrate people into Europe" to get blades and sharper flakes, which appeared as "the Levallois reduction" technique starting around 290,000 years ago—long before any modern men, and indicating, along with other evidence, no major radiation of African tools at the time of Berkeley's Eve but rather a previous migration of *erectus*.

When the "modern" tools came, they may have been more a variation than a revolution: bone spear points and specialized flint scrapers and gravers, in addition to the transitional blades, which probably did make life substantially easier and had more cutting edge. Some Neandertals, however, had started using Upper Paleolithic (or at least a kind known as "Chatelperronian") technology. In fact, it has been suggested that the Aurignacian culture associated with modern *sapiens* was just an extension of the Mousterian culture, and the interstratification of Neandertal and more modern cultures has suggested to others that there were no inhibitions between Neandertals and early moderns as far as personal interactions like mating.

If there was no clear evidence of a sudden, vastly superior technology in Europe, and if the European Neandertals indeed had been on the way to physical modernity, as demonstrated by transitional traits, then, in order for the Berkeley theory to hold up, there must have been some other advantage that allowed Berkeley's hypothetical Africans—the clan of Eve—to displace Neandertals.

As Lucy discoverer Donald Johanson wondered skeptically:

"What would have been coming out of Africa as *Homo sapiens* 200,000 years ago—what would it have possessed as an adaptation—that would have allowed it to completely displace and bring about the extinction of these *erectus* populations?"

With questions like that piling up, and the debate swirling faster than ever around the *Nature* paper, Allan Wilson decided to come out of seclusion and provide a new theory that would give Eve's progeny just the revolutionary and evolutionary edge it needed—and without a holocaust. The reclusive general, his troops increasingly bloodied, Cann getting fatigued, was finally stepping out from behind the barracks and marching to the front with a bullhorn.

He was a distinguished-looking man, yet with hard eyes, the guru's stare; and with his long white hair he bore no few resemblances to the maestro Sergiu Celibidache.

There was a great air of dénouement.

What, Dr. Wilson, was the elusive and mysterious evolutionary advantage that had dispelled everyone from Neandertals to Peking Man?

18

SHE COULD TALK. Or at least her progeny could. According to Wilson, Eve's band was the first to speak. Besides developing a slimmer body, a higher forehead, and less at the brow, the modern Africans had been the first to develop language. They talked their way out of Africa and into the Middle East, and sideways through Europe and Asia: unprecedented chatterboxes. "The Origin of Speeches," said a front-page headline in *USA Today.* "Scientist: Woman had the 1st word."

"Tongues will be wagging over this theory," said the article, on January 16, 1989, with certain prescience. "A biochemist thinks our capacity for language may have started with a genetic mutation that occurred in an African woman 200,000 years ago. And when her descendants colonized their neighbors to the north, language may have given them the edge over mute Europeans and Asians."

Wilson himself commented: "We think that we've found a mother who lived roughly 200,000 years ago and that she was one among many mothers living at that time—we think many thousands at least—and that she had two daughters, at least, one giving rise to Lineage I, which largely stayed in Africa, and then Lineage II, which includes many Africans and all the non-Africans." And in one of those two lineages: the first to really speak.

It was no joke. The Berkeley group needed a good explanation for how a small band of Africans could have stolen the world from Neandertals, archaics, and descendants of *erectus* in Asia, and Wilson, at a meeting of the American Association for the Advancement of Science (AAAS) in San Francisco, was there to provide it. "In his lecture to AAAS," said a blaring press release from Berkeley, "[Wilson] cited new support for this view and offered new evidence that modern humans—*Homo sapiens*—emerged only in Africa and may have swept aside other species in Asia and Europe."

Women, not men, may have passed on to their children the key trait that distinguished the first modern humans, explained the press release, adding that what Wilson was now showing was "how a maternally inherited ability to use language or perform some other vital skill could have allowed modern humans to replace more 'archaic' species about 100,000 years ago *with no intermixing*" (author's emphasis).

Neandertals and archaics were not only crude, said Wilson, but also linguistically deaf and mute. Eve, meanwhile, was ready to stand in for Johnny Carson.

Most stunning was the assertion that the ability to speak may have come from a mitochondrial (instead of an infinitely more likely nuclear) DNA mutation. What Wilson was thinking about was a mitochondrial mutation which, he hypothesized, may have caused more oxygen to go into the Africans' brains. That was what the mitochondrion was all about: oxygen and energy production. A brain with more oxygen and energy may have had the memory circuits that led to speech.

But this was really stretching it, and just about nobody believed it for a moment. How wondrous was the mitochondrial genome: not only did it have unique quirks that allowed it to be traced back maternally, a mutation rate that ticked through the millennia like a clock, and intricate patterning that nicely planted twigs in the genealogical tree of mankind, but it also contained, most wondrously, surprisingly, and indeed momentously of all, a mutation that had allowed Eve and her Africans to take away the world from the "mute" descendants of Neandertals, Ngandong, and Peking Man.

Though the nucleus had a *billion* or a few billion base-pairs, while the whole mitochondrion had only 16,500, and a mere 37 genes, it was in the mitochondrion's DNA—not really known, hitherto, for causing much but thirteen of its own proteins along with its ribonucleotides—where, we were now being told, the mutation had occurred that brought mankind where it is today.

Mitochondrial DNA, that genetic material which had first won Wilson's affection as an evolutionary "clock," was no longer just quirky genetic material that served as an evolutionary time-piece but instead, coincidentally, was now the key player in mod-ern evolution.

Wilson was doing at least three things with the latest and most controversial theory of his career: he was throwing his pursu-ers a diverting scent, daring, mocking them to disprove an unprov-able hypothesis; he was baiting the media, which he loved to play; and he was stretching his credibility to the danger point, calling all his previous assumptions into serious new question. "If scien-tists balk at such talk, it won't be the first time Wilson has sparked heated dialogue," noted the middle of the USA Today article with admirable understatement.

Wilson the gremlin. That was Cann's impression: he was playing catch-me-if-you-can. The mischievousness of Sarich had rubbed off, or maybe Wilson had been the more mischievous one to begin with. He didn't actually say, in his lecture, that Eve herself could talk, but it was strong stuff nonetheless, more speculating than Wilson had done since hypothesizing that Australia was the origin for modern humans, and a nervous Cann, for one, thought or hoped he was just teasing the press. It was the only explanation one could fathom without pointing to serious contradictions and really more serious questions than that. After all, it was Wilson who'd told me he dealt in statistics while he let everyone else stew in their opinions—coming up now with a theory that, by the kindest reckoning, was itself statistically mute and the height of scientific conjecture.

He was the man who tortured paleoanthropologists for their assumptions and subjectivity (which were legion), still taunt-ing Simons and Pilbeam for thinking the ape-human split might have been 15 million or more years old (instead of 5), now telling the world that the same little piece of DNA which through odd and fortuitous properties had allowed a new view of evolution—a piece of genetic material composing only ⅓,₀₀₀ of the human genes, chiefly to do with cell respiration, taking a big back seat to the DNA in the nucleus, 37 genes versus the nucleus's 100,000— announcing that mitochondrial DNA, in Eve's line, no less, had decided who could talk and who couldn't.

If it was a game to Wilson, or just trying to beat back the enemy, or simply that he was energized with a brilliant new idea, he may have underestimated the ramifications down the line. The pronouncement was made at the annual convention of the AAAS, an entirely serious affair: there were people talking about AIDS, genetic engineering, and the "greenhouse effect." In the shadow of those concerns no issue of Neandertals versus *Homo sapiens sapiens* was really up to snuff, but Wilson seemed almost whimsical about what he said and what the media would transmit about the lecture. "Perhaps, Wilson suggested, a single mutation in the maternally inherited mitochondrial DNA accounted for the African group's superior abilities," emphasized Berkeley's public relations office, to make sure everyone got the point—and the right headlines.

While the *New York Times* didn't pay any attention, the story was making it on the wire services, which was apparently just what Wilson was seeking. Though, as a reclusive guru, he was supposed to hate publicity, the press release Berkeley issued was five pages in length and emphasized that Wilson was the one whose research group had "startled colleagues" two years before with the Eve hypothesis.

It cast a pall over the Eve hypothesis, begging a second look at a theory that had rattled all of evolutionary research now for two years, that had become the new orthodoxy, that *Newsweek* had generously accorded its cover page, but which had been suffering of late and was starting to look, at the very least, a little wobbly. The idea Wilson was now promoting seemed to possess what Cann called "logical inconsistencies." It was difficult not to be taken aback. It was reminiscent of another theory put forward the previous June by a Berkeley anthropologist who figured the myth of Noah's Ark evolved from male envy of female reproductive capabilities (the Flood being a symbol for female amniotic fluid!).

As for the ability of the ancient Africans to talk, and how they had taken over the world, eliminating the Eurasian genes, Wilson had observed (said the Berkeley press release) "that when one group invades or conquers another, mixing between the invading group's men and the resident group's women occurs, but not the reverse. Children of the mix would inherit the resident species' mitochondrial DNA, so they would lack the advanced ability [to talk]. They would be likely to share the resident population's fate: extinction. Only children of the African women would possess the new, vital skill."

In essence, then, what Wilson was using to show that it was a mitochondrial mutation and not a seemingly much more likely nuclear mutation that had triggered the suddenness of language was the very thing he had said (and repeated in the same press release) had *not* occurred between Eve's throngs and the hopeless archaics: admixture. And if there was conquering—if only the invading men and conquered women intermixed—this implied just what Wolpoff had said it implied: a violent takeover, warfare. Meanwhile the much ballyhooed mitochondrial DNA whose portrayal in the literature had the effect of presenting it as nearly an aberrant slice of neutral DNA with little to do but keep good molecular time and make ribonucleotides and its own proteins (and not even all its own proteins) now contained one of history's most consequential mutations.

Though neither Cann nor anyone else doubted that language played a key role in human development—she'd proposed this herself—and though she thought Wilson had "some real ideas" (perhaps it *was* possible that mutations in the mitochondria could affect large portions of brain tissue), she couldn't see how a mitochondrial mutation would affect something like verbal storage of information, and her impression was that her former mentor, the man who had encouraged her to pursue the Eve research to begin with and had been so crucial in formulating an evolutionary hypothesis that was now entering college texts, had "gone off the deep end" with his new notion.

Erik Trinkaus and other scientists were less restrained. Trinkaus called it "an embarrassment to science, and I hope you quote me on that." British paleontologist Chris Stringer, the Wil-

son lab's greatest supporter among the stones-and-bones contingent, chuckled nervously and skeptically into the phone when he
heard of it. "It's very weird," he sighed. Geoffrey Pope, a University of Illinois paleoanthropologist who specialized in Chinese
remnants and had gone to Berkeley, a friend of Cann's but an
intellectual ally of Wolpoff's, agreed that Wilson had "gone
around the bend" with his mitochondrial DNA. It was the pronouncement, the "Talking Eve" hypothesis, which had led Randall Susman of Stony Brook to declare that this was the sort of
thing that gave the field a bad name. Wilson was "trying to enter-
tain people," commented Morris Goodman, whose work with
protein had spurred Wilson in the 1960s. When Wolpoff heard
about Wilson's explanation of how Eve's clan could have taken
over the Old World from archaics and Neandertals (without the
archaics leaving telltale genes) the anthropologist had roared with
unrehearsed laughter. "He's flipped out!"

The assessments were almost universally harsh. Sarich,
who still sat at a desk in the Wilson laboratory, and was quite
familiar himself with propounding radical theories, thought the
concept was "nonsense," had argued with Wilson that proposing
any such thing would be "harmful," and said hardly anyone in the
laboratory was conversing about it, deciding that it was a "temporary aberration."

"Let's not even talk about that," Sarich begged off.

"Nonsense" seemed to be the most common reaction, or
"absurd." Clark Howell fairly ran away from commenting on it,
urging an inquiring journalist to seek guidance elsewhere. One of
the kinder reactions had come from Berkeley's Tim White, who
smiled that the hypothesis was "data-free." Others dismissed it—
no pun intended—as not worth talking about.

Needless to say, the "Talking Eve" idea flew in the face of
many paleoanthropological beliefs, chiefly the feeling that perhaps
as long ago as a couple million years man-apes had begun a system
of complex communication. Even vervet monkeys had a chitter for
snakes, and a r-r-raup for birds of prey, and Washoe the chimp had
understood 350 hand signals by the age of five. Gorillas could be
trained to just about order at McDonald's, and though paleoanthropology happened to be going through another period where

the language capabilities of *erectus* and Neandertals were again placed in doubt, there were many scientists who felt confident that *erectus* or surely Neandertals could have uttered a few words, if not whole rudimentary sentences. All those tools: cavemen had to have referred to them differentially, with some kind of conceptual template or labels. And one Neandertal in Israel (see the Notes for Chapter 20) was found with a bone structure near the larynx that is crucial to language and was identical to that of modern humans.

Whatever the problems with Wilson's latest insights, his critics needed to remember, in the midst of this "aberration," that the Berkeley biochemist was a man of no mean accomplishment. From a modest start on New Zealand's South Island Allan Wilson had progressed to a level of scientific achievement that was in fact the stuff of any academic's dreams. It was certainly not a career that could have been built on absurdity and nonsense. The language was harsh, as often language in anthropology was harsh, but no one could truly have meant he had flipped out. He had simply dared to combine two volatile characteristics: creativity and chutzpah. It was kind of endearing and raised a question. Did people call him absurd simply because they didn't like him, or out of jealousy? Even though he was all but anonymous to the public at large—another scientist hidden on some campus—he was getting very well known within the small, stuffy world of evolutionary studies. In the literature it was now often Leakey, Wolpoff, and Cann/Wilson who got the citations. To a cloistered academic community, with its collective nose stuck in the technical journals, *that* seems like fame. And fame breeds jealousy. Not only was Wilson an elected member of the Royal Society of London and the American Academy of Arts and Sciences, but he had also served as a reviewer for *Nature, Science,* and at least thirteen other journals.

Moreover, his Berkeley lab had produced a torrent of technical papers with far less flamboyant conclusions and perhaps important ones. There were more than 300 of them, by Wilson's count, and they ranged from studies of salmonid mitochondrial DNA to cow lysozymes. He wasn't always looking for Eve. And none of this, nor his theorizing, had come cheaply. Between 1973 and the beginning of 1989 the NSF alone had supplied Wilson with $1,994,378 in grants (according to figures supplied under a Freedom of Information Act request), and the National Institute of General Medical Sciences, but one division of the NIH, had offered Wilson, as principal investigator on grant applications, another $2,349,780 between 1972 and the onset of 1989, the vast majority of it for studies of "organismal differences," which implied a lot of evolution research. So the total for these two agencies alone came to $4.34 million, and did not take into account other agencies, private grants, and what taxpayers in the state of California had contributed.

Could anyone who obtained money like that be "absurd" or full of "nonsense"?

While the NSF was a likely reservoir of such financial support, the institutes of health were a bit more puzzling. When I asked Wilson the medical applications of his research, he responded, "Indirectly, there are many. Time does not permit a discourse on this subject."

At the NIH, there was shuffling and edginess when the Eve work was mentioned. "Whenever I see *that* in the paper I say to myself, 'We didn't fund that,' " one official tried to reassure herself (even though Wilson listed NIH in his Eve papers), while Wilson's grant administrator, Joye Jones, thought the language hypothesis "weird at best." When asked about the medical benefits coming from Wilson's lab she replied, "A direct benefit of Allan Wilson's work I'm not sure I can come up with." (But she quickly added that there were indeed "indirect" benefits of his other work, including new knowledge of amino acids, helpful information that others, using his techniques, had unearthed about fungal infections, and always the chance that his research would turn up a "serendipitous" discovery.)

It was Wilson's belief that "much of the furor about what we've written has come from the fact that most biologists did not have a firm grasp of population biology, not even basic principles." At the science convocation he hinted that it was the press that sensationalized his ideas, predicting so rightly that the media would pick up on his language theme. "They'll dwell on the hypothesis, perhaps unnecessarily so, that the invention of language in Africa may have been the factor that allowed this group to expand throughout the world without genetic intermixing with the resident population and that the genetic basis of this ability may include a key mutation in mitochondrial DNA," he had said, as if he didn't really want the press to dwell on it despite Berkeley's five-page press release.

Wilson's voice was somewhat lilting, with the trace of a New Zealand accent. When he'd gotten to the part about the hypothetical mitochondrial DNA mutation, it had lowered a bit and showed a little hesitation. But Wilson went through with it, and the lecture in San Francisco was, if nothing else, an act of academic bravura. For besides introducing an idea that was so speculative the vast majority of scientists would have avoided such a thing like the plague, Wilson was doing his tap dance on the limb of a tree that other scientists—both geneticists and anthropologists—were sawing at the trunk. At issue was the very foundation of his premise for both the ape and Eve research: that there was a reliable molecular clock.

For anything to function as a clock, there were two basic standards it had to meet. It had to be relatively constant (no one liked a clock that started and stopped), and its units of measurement—its hours and minutes, the mutation rate—had to be universally accepted. What was an hour to one timekeeper couldn't be half an hour to someone else. In other words, the rate at which the hands of a molecular clock moved was something that everyone would have to agree upon.

Yet there was anything but agreement that Berkeley's 2 to 4 percent rate of mutations was correct, and even Cann was starting to say that the rate—double Wes Brown's original estimate—was too fast. This was key because the slower the rate, the further back in time was any "Eve," and the more likely would it become that she wasn't so much an Eve as a mute, stubby ape-woman.

Brown, the very man who'd started the first human sampling and had brought the mitochondrial technology to Wilson's lab, had continued his mitochondrial work in Michigan and still believed Berkeley's rate was "twofold" off. In other words, every half an hour of evolutionary time that Wilson had logged had been an hour to Wes. Such discrepancies add up over a period of one million years. Brown (who, as we've seen, had long been disturbed by how quick Wilson was to form a public hypothesis and who was very disturbed still, especially over matters like the language speculation, which he said "sounds to me like another one of Allan's off-the-wall bullshit speculations, which he's made many, many, many times throughout his career") called the *Nature* paper overhyped and unbalanced.

If the mutation rate was, as Brown suggested, 1 to 2 percent instead of 2 to 4 percent, that would double the date for Eden and the existence of a hypothetical Eve could then be pushed back twice as far in time—to 285,000 to 570,000 years, which would have meant what Wilson's lab had been documenting was perhaps not so much an archaic soon to give rise to modern humans but one of the oldest archaics like Rhodesian Man or Bodo that probably came long before the transition to *Homo sapiens sapiens.* For all anyone really knew, it may have even been pointing to a late *erectus!*

The lower the rate, and thus the further back in time the estimates went, the more probable it became, especially if it got back to between 500,000 and a million years ago, that Berkeley had been tracking not the rise of modern man, which is what had gained the lab its notoriety, but instead the radiation of *erectus* from Africa and the already well known establishment of *erectus* descendants of Peking Man himself around the world—in effect, a genetic confirmation of Wolpoff's theory.

"Several studies by American, European, Japanese, Israeli, and Chinese investigators show that mtDNA morphs and types specific to several populations outside Africa give estimated divergence times much older than 180,000 years ago," said James Spuhler of the Los Alamos National Laboratory, whose training had come at Oxford and Harvard, and who had chaired the anthropology department at the University of Michigan. "Cann, Stoneking, and Wilson are simply mistaken; the world distribution of mtDNA types gives stronger support to the multiregional than to the single region model for the origin of *Homo sapiens.*"

These were criticisms which were coming even before Wilson had damaged the Eve theory with his language speculation. One of the most widely cited geneticists studying such issues, Masatoshi Nei of the University of Texas, whom the Wilson lab itself cited in its references, told me his estimated rate for a single lineage (as opposed to a rate calculated for a double lineage, which is what Berkeley used) was 0.7 percent per million years. Double that to account for a double lineage, and one came up with 1.4 percent, which was still substantially less than the Berkeley range. "Essentially my estimate is half of their estimate," says Nei. "So in my view the woman can be *Homo erectus.*"

(When I asked Nei, who himself had quite an investment in mitochondrial DNA research, the one question he'd most like to pose to Wilson, he replied, "I would want to know if he still really believes this one piece of DNA can tell the history of *Homo sapiens.*")

Then there was the issue of constancy: Did a "molecular clock" keep good time? While Wilson had warned that the "clock" functioned more like a radiometric one (with small fits and starts that averaged out) than a metronomic clock, his group had emphasized that overall the mutational rates, at least in mammals, was fairly constant, and Wilson had published a paper which mentioned that the 2 percent rate applied to everything from Hawaiian fruitflies (*Drosophila*) to apes. Elsewhere he had said there was remarkable constancy of mutation rates in organisms as disparate as plants and bacteria. It was certainly constant enough, if Wilson was right, to determine the age of ape divergences that went back 5 million years.

However, another former worker in the Wilson laboratory, Rodney L. Honeycutt, whose own wife had donated a placenta for the Eve study, felt "their range of error on their time estimates is as great as their average estimates of time themselves, and there's this no-good calibration. They're assuming that there's a 2 percent per million years divergence that goes across all mammals, and they really don't have any good data on primates to suggest that."

Honeycutt, who was at Harvard when I reached him (since moved to Texas) and was working actively with mitochondrial DNA, said his own lab was obtaining evidence that there had been erratic changes in the rate of primate mutations, a major shift in amino acid replacements during one especially noteworthy period of time. The whole primate lineage, in fact, had undergone, he said, a "drastic rate increase" in a particular mitochondrial gene compared to other mammals, and the same thing, says Honeycutt, could have occurred in man.

Brown, meanwhile, had recently conducted research sharply questioning the constancy of molecular rates, especially the idea that mutations in mitochondrial DNA always occurred faster than in the nucleus. On October 10, 1986, in *Science*, he'd published a study on sea urchins entitled "Nuclear and Mitochondrial DNA Comparisons Reveal Extreme Variation in the Molecular Clock." In the summary Brown said his analyses provided "a robust rejection of a generalized molecular clock hypothesis of DNA evolution."

"In sea urchins data indicates mitochondrial DNA changes if anything more slowly than nuclear DNA does," says Brown's rising voice. "And in some Drosophila that also appears to be the case. In some cases mitochondrial DNA is changing much more rapidly than nuclear, in other cases it's the same rate or even slower. Well, that's a five- or tenfold rate difference. Where's the clock? If that were your watch, you'd throw it away."

That Brown was so candidly at odds with the Wilson lab's foundational rate—the basis of its highly publicized dates—was at best discomfiting. One could hypothesize that there were the forces of personality at work. Though Brown expressed no upset that his name had not been tacked onto the *Nature* paper despite his earlier contributions, his relationship with Wilson by his own reckoning had hardly been the picture of congeniality. "I have to admit I was a little upset that Allan has been loathe to cite that original study I did," says Brown. "He manages to cite it in the most obscure way possible when he [does] cite it."

But Brown did not seem to be playing ego games. He was widely respected as a conservative empiricist—which may have been all it took to get mad—and there were still others like Morris Goodman who believed there had been, as many feared, a rate slowdown in man. "We find sometimes systematic differences in rates," says Goodman, whose work was now with nuclear DNA but who felt what applied to that type of DNA also applied broadly to that of the mitochondrion. "I suggested to Wilson that the chances are when he fills out that tree with non-humans from other branches of the primate family such as Old World monkeys,

New World monkeys, and so on, he'll find that the real fast rate
is in the early primates and then it becomes slower in the hominoids."

Wilson had little time for what Goodman was saying ("I'm
pretty intolerant of the idea of a slowdown"). But Goodman
wasn't the only one implying this. In March of 1987 Wen-Hsiung
and Masako Tanimura of the University of Texas, also studying
nuclear DNA, published a paper whose headline said it bluntly:
"The molecular clock runs more slowly in man than in apes and
monkeys." In all comparisons between an ape lineage and the
human lineage, they said, "there is no case in which the human
lineage has evolved faster. This is true for both nuclear and mitochondrial sequences. When all the nuclear sequences are considered together, the rates in the orangutan, gorilla and chimpanzee
lineages are, respectively, 1.3, 1.9 and 1.6 times faster than the rate
in the human lineage."

Still, there was nothing yet to wipe the molecular clock
entirely from the screen, and though it was not as constant as
originally portrayed, and showed all kinds of quirks in lower animals, Brown himself felt it was still generally regular and of some
evolutionary use. Because molecules always change, they remained
a timepiece, albeit a crude timepiece that was imperfect and had
to be used with much more caution than what Brown had wit-
nessed thus far. It was still likely valid between closely related taxa,
at least for now.

But there remained the issue of rate, and also arguments
that a midpoint rooting of the DNA tree and the overall method
of parsimony gave misleading results. From the University of California at Davis had come geneticist John H. Gillespie even before
the *Nature* study with his opinion, which he still maintains, that
"analyses of protein sequence data may have been distorted by
unrecognized biases that have led to gross underestimates of the
variability in the rates of evolution." These biases, said Gillespie,
meant that "the inferred dynamics of molecular evolution appear
to be much more erratic than suggested both by neutral allele
models and by the molecular clock hypothesis."

Then there was Roy Britten from the California Institute
of Technology, with indications that DNA rates varied not only

between lineages, but at different locations in a single genome. "I just got back from a meeting in Lake Tahoe of molecular evolutionists, and the consensus I gathered was that there was no 'Eve,' " says Britten, who was also involved in the ongoing Sibley controversy. "If someone had mentioned the language hypothesis on the stage, I'm sure it would have gotten the biggest laugh of the whole time."

Wilson suffered little patience with objections about the rate he was using. "I don't think they've read Stoneking, et al., 1986," was his response to Brown and Honeycutt, referring to his and Stoneking's paper published in a Cold Spring Harbor volume in which Wilson explained how the rate was calculated at least in part by looking at the New Guinea population. As for the other criticisms, "Who's Spuhler?" was about Wilson's only response.

Brown shot right back. "I don't think *he's* read others' papers except superficially," he said, adding, "Allan is soft-spoken but I wouldn't characterize him as shy or cautious. One point of conflict Allan and I had in the lab—and it was an open conflict—is that I feel he is often much too fast and loose. And that he's very quick to hypothesize things and talk about them and put them into print even when, as far as I'm concerned, there's insufficient data really to do that."

With this kind of blood in the water—Wilson's own colleagues not so much sniping as throwing grenades—it was no surprise that Milford Wolpoff would make an appearance. In coming forward with the "Talking Eve" theory, Wilson had walked, in fact, right into Wolpoff's sights.

Wolpoff believed any mutation in mitochondrial DNA that had an evolutionary impact would provoke the forces of natural selection to act upon that molecule in such a way as to muck up its use as a clock. For one thing it would cause the clock to be reset every time such a mutation allowed one lineage of DNA to survive beyond another kind that was older. "Natural selection," he said, "kills the molecular clock."

Not every geneticist thought that whether natural selection operated on mitochondrial DNA was especially relevant ("Of course it's under selection," Wilson huffed), but this wasn't all that Wolpoff had to offer. He pointed out that in Europe there were

the mice Wilson himself had reported that possessed the mitochondrial DNA of one species but the nuclear DNA of another.
It appeared, in other words, that mitochondrial DNA could per-
vade a new territory while the much more relevant nuclear DNA—
which determined what a person or creature looked and acted
like—was stopped near the border. "Therefore," said Wolpoff,
"the pattern of today may simply show random survivorship from
what earlier was a single very common, widespread mitochondrial
lineage."

Taking this just a short step further, it became easier to
imagine that while an African woman's mitochondrial DNA may
have survived through the millennia by simple chance, this single
compartment of genetic material could have existed in a cell with
nuclear DNA which contained her genes but also genes from a line
descending from European and Asian archaics. Though, by simple
luck, or what is called "stochastic loss," only the one woman's
mitochondrial DNA survived, hidden in our nuclear DNA may be
thousands of old European and Asian genes that simply cannot be
detected by current methods—perhaps derived from Neandertals
or Peking Man himself.

Such a possibility recalled the analogy of vanishing surnames. It should be remembered that initially mitochondrial DNA
researchers were rather convinced that there had been a tiny
population—or "bottleneck"—which had given rise to modern
man, and that such a population may even have been the Adam
and Eve everyone (especially Wilson) hoped for: a single mating
couple. Although still possible, the idea of such a literal genetic
Eve all but disappeared once John Avise of Georgia had argued
that mitochondrial DNA moved down through the ages like family
names. If a couple has only daughters, the father's surname does
not make it to the next generation even if his nuclear genes sure
do. With mitochondrial DNA the situation is analogous but reversed: in each generation certain women will produce no daughters, and thus will contribute no mitochondrial DNA to the next
generation although their much more important nuclear genes will
be passed on. Thus, because all current mitochondrial DNA can
be traced back to one woman hardly means she was the only
woman around. The mitochondrial DNA that survived down

through the ages represented more a "Lucky Mom"—if only one mom it was—than Adam's fairest.

And because we have her mitochondrial genes doesn't mean we have more than an exquisitely small fraction of her nuclear genes. That is why she has been referred to, time and again, as a "mitochondrial Eve" and not a real one.

But there are those like Wolpoff who argue that the mito- chondrial Eve may not have lived anywhere near the point of modern human origin and therefore is not even a mitochondrial "Eve." Some analyses of mitochondrial variation seem to have shown that such variation provides few clues as to precise geo- graphic origins of an Eve. One widely quoted study of fruitflies grants an idea of the past variables we cannot yet ascertain. "Today's world population of *Drosophila subobscura* consists of many millions of individuals," said the study, published during 1986 in the *Proceedings of the National Academy of Sciences.* "It might well be the case that, a few hundred thousand years hence, all *D. subobscura* flies have mtDNAs derived from morph I. That would not mean that the mtDNA of the descendants derives only from one *D. subobscura* currently living—morph I is found in 44 percent of the living populations."

While it was suggested that a group of women with identi- cal mitochondrial DNA types—and not a single ancestor—may be the source of genetic diversity in modern humans, Stoneking and Cann argued back in their Cambridge paper that "this hypotheti- cal group of women must be descended from a single common ancestor in a preceding generation. It is not biologically feasible to have multiple lines of descent without a common ancestor."

This was all part of the convoluted technicality you had yet to suffer to get to the real Eve. (Those beckoning caves!) Also, outright nastiness was entering the picture. Tim White, who had been one of Wolpoff's students but was now a bitter foe of the Michigan morphologist, thought the mitochondrial DNA was something everyone had grasped much too soon, but he wasn't about to agree with Wolpoff's arguments. Were there any people still living who, by Neandertal features, could prove that archaics contributed to the modern gene pool? "Not even Milford has been able to argue convincingly that there's a living group of Neandertals around," replied White sarcastically. "[Though] some of my colleagues have suggested that he approximates it."

Of more immediate importance than such *ad hominem* remarks were the questions about time of origin and experimental procedures. Besides accepting a rate that was too high and thus estimating time depths that were too shallow, said Spuhler, the use of a polyacrylamide gel may have distorted interpretations of variances between populations. Spuhler got in a dig saying Wilson and Sarich's 5 million for the chimp divergence from man was two times too late, while their rival Elwyn Simons, whom they had castigated with little mercy, had been off by not much more: a factor of three. Indeed, looked at from one extreme, if Spuhler's estimate of 10 million for the ape-human divergence was the correct date, those paleoanthropologists who'd believed the divergence was 15 million years ago (and believed *Ramapithecus* was a hominid) had been no further from the truth than Sarich and Wilson. Moreover, Spuhler said estimates for evolutionary rates among New Guineans, Australians, and American Indians were not trustworthy "because all three regions were peopled by two or more major migrations rather than being settled once and isolated thereafter."

There were other potentially confounding factors as well. One scientist believed genetic material could piggyback on viruses moving from species to species, making them appear more closely related than they actually were. Could viruses also have mucked up human relationships, to one degree or another?

These questions, criticisms, and attacks, it should be realized, came at the same time Sarich, Marks, and Wilson were waging war with Sibley and Ahlquist. There was gunfire everywhere. In 1988, the journal *Immunogenetics* ran the review of a new textbook, *Molecular Biology of the Gene,* that castigated this textbook for including three studies or notions—including the work with hybrid mice—that Wilson had been involved with. "Three sections in a textbook, three false alarms, and just for the sake of sensationalism," complained the reviewer, Jan Klein. " 'Wormy Mice in a Hybrid Zone,' 'Mitochondrial DNA Invades House Mice,' 'Eve Did Exist!' What newspaper headlines these would make—and in fact, they did! It is bad enough when journalists exaggerate, twist the truth, and make up sensational stories; it is much worse when irresponsible scientists make unfounded claims and play into the hands of unscrupulous journalists; but it is totally unacceptable to make sensationalism part of a textbook and teach students things that simply are not true."

Undaunted and in a measured tone, Wilson had responded that he partly agreed with Klein. "Eve," he said, was "a poor term for use for the common ancestor of all maternal lineages in a species. In publications from my laboratory, the term Eve has always been avoided."

It was a short while later that Wilson gave his plenary lecture at the AAAS. His talk was entitled "The Search for Eve."

From way out in left field during that same conference had come a truly troubling development for evolutionary biology in general and geneticists like Wilson in particular. Citing genetic evidence as a basis for his conclusions, a psychologist from the University of Western Ontario in Canada, Philippe Rushton, reported that if races had begun to diverge 110,000 years ago, with blacks emerging first, this indicated, along with other ordered differences among the races that he had noted, that blacks were the least intelligent of the races and had the highest sex drive. Whites, who he said emerged second, had more intelligence but less of a sex drive, while Orientals were of the highest intelligence and concomitantly the lowest sex drive. It was an ironic reversal of Coon, who had argued that blacks lagged behind because they had developed last. It was also ironic because where Coon had been on the side of continuity, Rushton was clearly fond of the molecular viewpoint.

That was admittedly just an unworthy diversion, but on a technical and more responsible level most nonplussing of all was a paper entitled "Origin and Differentiation of Human Mitochondrial DNA" that appeared early in 1989 in the *American Journal of Human Genetics.* The authors, Laurent Excoffier and André Langaney of the Laboratoire de Génétique et Biométrie at the University of Geneva, took a look at the human mitochondrial results from several laboratories, which represented 1,064 individuals who had been sampled up to that time (as opposed to Berkeley's 147). They decided Berkeley's genealogical tree had been biased by "topological errors" and therefore that the Eve hypothesis "cannot be supported by these data."

The paper caused a commotion in the Wilson lab. They were trying to decide whether to respond or ignore it. That said something about its importance. It had to have been upsetting to anyone with an emotional or professional investment in the Eve idea. According to Excoffier and Langaney, African populations appeared to have "very differentiated mtDNA types, most of which have appeared only recently. *They are not, therefore, likely heirs of a population from which all others have diverged"* (author's emphasis).

Indeed, Excoffier and Langaney, in comparing results from Berkeley and other laboratories, found, under one model, that some Caucasian and Oriental samples appeared more differentiated than African samples "and could be derived from older ancestral populations." The finding of two DNA types in an Israeli Arab sample indicated to Excoffier and Langaney that present African populations "may have originated from a common and genetically extra-African population existing before the continental splits." They found suggestions that "a selective mechanism favoring certain types or eliminating others is at work." The time necessary to produce the present molecular diversity, they said, "largely exceeds 100,000 years" and may have been in excess of 400,000 years. In short there were indications, in the accumulating mass of collective mitochondrial data from around the world—from laboratories in Israel and Japan and the United States—that Caucasoid populations "could be the closest to an ancestral population from which all other continental groups would have diverged."

Now it looked like Eve might have been milk-skinned—or lighter-skinned—after all. But others said she may have had slanty eyes. Indeed, one of the geneticists in a lab Excoffier and Langaney had evaluated, working with the same basic tools as the Berkeley group and sampling more people than Wilson's lab (including authentic Africans instead of American blacks), had come to the preliminary conclusion, it seems, that Eve was not only "extra-African" but was from the territory of Peking Man.

19

THAT LAB WAS THE ONE occupied by Stanford geneticist Doug Wallace, who was now in Atlanta. Wallace had not sliced up his DNA to the fine parts Berkeley had, but his human sampling was more extensive and his thinking Asia brought him into no little conflict with Cann, et al. That Wallace could lean in the direction of an entirely different continent suggested that DNA could be interpreted with the same subjectivity as fossils, showing that genetics and paleoanthropology had something in common after all.

Douglas C. Wallace was a short and energetic man whose days were as hectic as Wilson's but with far less exotica and controversy. He'd been involved with human mitochondrial DNA before it had become fashionable at Berkeley (his fascination with the genetic molecule harkening back to the early days of Wes Brown), and it had been Wallace who had indisputably proven a number of mitochondrial DNA's peculiar tendencies, such as the fact that in humans it is inherited only from our mothers.

Berkeley was the competition, and there was some bad blood between the Wilson and Wallace labs. On the joke board where Wallace was now, at Emory University, was a newspaper clip about the new Eve hypothesis Wilson had announced. A notation next to it asked: "And what language did she speak?"

Wallace agreed with a lot of things Berkeley was saying, such as the mutation rate and the fact that mitochondrial DNA indeed indicated a single population expanding around the world.

This flew in the face of Wolpoff, but he didn't take the matter as a war with Wolpoff and there was more antagonism with Cann and Wilson than with the paleoanthropologist from Michigan. Wallace had debated Wolpoff on the radio and found him enjoyable and witty.

On the other hand, turning to the West Coast, Wallace criticized Berkeley's use of American blacks to fathom African origins and described Cann, et al., as having fallen "into a classic population genetics design error." This wasn't the kind of overt shouting that occurred between paleoanthropologists, but it wasn't hard to read between the lines. "I would not have done their study their way," says Wallace. "We did *not* do our study their way. We went first to aboriginal populations in collaboration with anthropologists who knew those populations, because we felt the worst thing that could happen in this field is to choose a population that had recent admixture from two populations. Your phylogenies would never make sense. The only way to do it is to go to Africa and define Africans as Africans."

Cann had admitted the problem with assuming that American blacks were representative of African DNA, but there was evidence now that it hadn't compromised their hypothesis (recent analyses of authentic Africans confirmed their tree), and she thought criticisms like those from Wallace's group might be more a matter of competition than scientific dialogue. "They thought we came out of nowhere, that we didn't have any right to work on mitochondria," she says. "I think it was really threatening—a lot of fear and a lot of suspicion. I know that he was threatened by the whole thing, and when we meet in person he is very personable but I don't trust him and I know he feels the same about me."

Born in Maryland, Wallace had gone to Stanford after schooling at Cornell and Yale. Where Wilson had a lot of chemist in him, Wallace had been specifically trained in genetics, coupling that with a degree in microbiology. His first faculty position had been at Stanford in 1976, teaching genetics to aspiring doctors and medical researchers, and it was at Stanford that Wallace had teamed up with Luigi Luca Cavalli-Sforza, who had been one of the very first to apply high-tech genetics directly to human anthropology. It was back in 1963, we might remember, that Cavalli-

Sforza had found that Asia seemed genetically quite separate from Europe and Africa, suggesting to some that Asia was the origin.

In 1981, six years before Berkeley's *Nature* paper, Wallace, Cavalli-Sforza, and five other researchers (see Notes) had collaborated on a paper for the *Proceedings of the National Academy of Sciences* in which they reported the analysis of 235 individuals from five ethnic groups, including Bantus and Bushmen from deep Africa. One difference observed between populations was a fragment pattern, or "morph," that was much more common in Orientals than it was in Africans. Known as "mtHpa I-1," the morph was present in 12.5 percent of Asians but only 4 percent of African Bantus. It was not detected in either Pygmies or Bushmen. This seemed important because mtHpa I-1 was also found in apes, indicating that it was an ancestral morph. Since Orientals had more of the morph than other groups, Asia was taken as the oldest place of human ancestry.

"The most likely source of the original mtDNA pattern for the formation of the human ethnic groups seems to be Orientals," wrote Wallace, et al., in 1981, adding: "these results suggest that formation of human ethnic groups took place in the last part of the Pleistocene, starting in Asia and radiating towards the other continents, probably within the last 50,000–100,000 years. An estimate of the divergence of human ethnic groups based on mtDNA variation has been made in the range of 10,000–50,000 years, correcting a previous estimate and in agreement with an earlier estimate of 25,000–100,000, based on classical genetic markers."

The earlier estimate Wallace's group was "correcting" had come from the Berkeley laboratory. Scientists were subtle in print, but that indicated the level of rivalry. Cann, et al., in turn, explained that their technique had higher resolution than Wallace's, and though the group of which Wallace was a part had come out with a larger human sampling before Berkeley, the Wilson lab had phrased its study in a far catchier way and had come out ahead on the public relations front. The *Nature* paper had grabbed more headlines and was now more widely quoted than Wallace's 1981 study. So it was a footrace. Currently they were locked in a battle to see who would get into print first with data on the origin of American Indians.

Aside from simple rivalry, Wallace and Wilson were two very different men. Where Wilson was off on the speech kick, Wallace, far more low-keyed, was talking about an actual mutation that seemed to cause a serious eye disorder known as Leber's hereditary optic neuropathy. The disease caused young people to lose their sight. That mitochondrial DNA might cause blindness didn't say much about its "neutrality," but it was important medical news, and Wallace figured mutations might also be responsible for diseases that affect skeletal muscle and the brain.

This is doubtless where some of Wilson's ideas on language came from. That a nucleotide change in the mitochondrion could have a neuropathic effect paved the way for speculation that a change in mitochondrial DNA could also have affected the brain enough to open the door to speech. It even gave some credence to such an idea. But Wallace felt there was no evidence whatsoever to support Wilson's "Talking Eve" hypothesis, and he was trying now to remain noncommittal about the place of man's origin, although he still thought Asia was a decent possibility. Others, he had noted, hypothesized an Asian origin based upon the presence of a type-C virus in both apes from Asia and human beings. George J. Todaro of the National Cancer Institute had found that human DNA has a pattern similar to the virogene DNA of orangutans and gibbons, which are from Asia, at the same time it is different from the African gorilla and chimp.

One could also cite a study in China at Fudan University in Shanghai, in which mitochondrial DNA from 273 individuals was analyzed and in which it was found that two of China's nationalities, Han and Hui, diverged from two others, Uigur and Kazak, some 237,000 years ago, which didn't fit well with a 200,000-year-old African Eve. The Fudan group too concluded that the common ancestor was Asian.

By 1989, Wallace had analyzed more than 800 human samples, but even back in his original study, he'd looked at 133 Africans—as opposed to Berkeley's two. That meant Wallace and his colleagues had seen nearly as many native Africans as the Berkeley group originally had seen of all ethnic groups combined.

Although his group had reached the preliminary conclusion of Asia, he liked to qualify that finding. "I have never come out and ardently said that the origin was either Africa or Asia," says Wallace, explaining that a second way of looking at his data—focusing upon the area of most divergence, which is Africa—led him, before the Berkeley conclusion, to also raise the possibility of an African origin. "The reason I've been ambiguous is that I don't know the answer. Only God knows. What we found is that the amount of genetic variation was greatest between the Bushmen and the other ethnic groups which include the Bantus, the Caucasians, American Indians, and Orientals. The Caucasians, American Indians, and Orientals, by contrast, were most closely related to each other. This implied that the origin of mitochondrial DNA occurred between the Bushmen and the Bantus and therefore was in Africa."

But there was the matter of morphs: mtHpa I-1 was in the orangutan but not in the Pygmy or Bushman. Orangutans had been indigenous to the Far East, not the sub-Sahara. One could tell Wallace still thought quite a bit about Asia. And if humans did originate there, Wallace figured, it might have been southeastern parts of that continent like Vietnam.

Still, Wallace hedged his bets. If you didn't, it was easy to get burned. In the 1960s, Cavalli-Sforza had taken the view of Europeans being closer to Africans than to Asians (implying an Asian origin) but in 1988 had reversed himself, declaring now that *Africans* seemed separate from the rest of the world, and that Europeans were genetically closer to Asians. This was in basic agreement with the other early and major researcher, Masatoshi Nei, who had started publishing divergence times back in 1974. Nei's data indicated that Negroids and the Caucasoid-Mongoloid group diverged from each other 110,000 years ago, whereas Caucasoid and Mongoloid groups didn't separate from each other until about 40,000 years ago. That is, Europeans and Asians had continued down the evolutionary path together after separating from the rootstock of Africans. Back then this was largely ignored by paleoanthropologists who thought modern man was only 25,000 years old.

Although Cavalli-Sforza's leaning on the place of origin

was now more in tune with Cann, et al., than Wallace, he was not necessarily in quite as much agreement on the issue of whether Neandertals and Asian *erectus* had contributed to the modern gene pool. "I have suggested to Allan [Wilson] that it's perfectly possible that the strange branch he has, leading to about one third of Africans, might perfectly well be Neandertal or earlier *sapiens* or who knows what, because he cannot say where it comes from, he cannot say that first branch is really African," he told me.

It was Cavalli-Sforza who, in August of 1988, came up with an important paper suggesting that a linking of groups based on linguistic affinity showed that linguistic families shared a striking correspondence with genetic clusters—in effect, that languages evolved along with genes, an idea that Wilson embraced with fervor and which apparently had helped inspire his "Talking Eve" theory.

But while Wilson listened closely to Cavalli-Sforza's dissertations on language, he and many others were ignoring certain other opinions that this venerable Stanford academic who had helped initiate some of the very first measurements of genetic distances had to offer them. "We find," says Cavalli-Sforza, "that we can detect a split in the history of humanity which is in agreement with evidence from both archeology and linguistic evolution. And we find that we do confirm now—reversing our earlier attitude—that the first split is between Africans and non-Africans." But, warned Cavalli-Sforza, "it's archeology that tells us Africa is the origin. No genetic data can tell you that. It can only show you differences but not where something arose."

While Africa had the most impressive and greatest number of fossil specimens, there were new suggestions—and suggestions only—that archaic human forms had been evolving into modern men on the Asian continent around the time they were doing the same over in Africa. That the great landmass north of the equator and stretching from the Black Sea down to the now separated islands in the South Pacific had been home to the first anatomically modern man is a possibility that not even the most ardent proponents of African fossils, including the Leakeys, are yet willing to dismiss. Asia has fully one third of the world's landmass and more than half its population.

It also has some of the most perplexing fossils. In addition to the Hexian and Dali remains, which seemed to bridge some of the gap between us and old Peking Man, researchers such as Geoffrey G. Pope, formerly of Berkeley and now at the University of Illinois, are currently studying two especially interesting fossils that Pope believes may be transitional forms that further close that gap, showing continuity from the ancient *erectus* like Peking Man to the present.

If indeed these fossils are representative of mankind stepping successfully toward or even across the threshold into modernity from Asian ancestors, the Berkeley hypothesis of a solely African origin for anatomically modern man would collapse on this ground alone. One of the specimens, found around 1983 at Golden Ox Mountain (also known as Jinniu Shan), a limestone hill that sticks out like a thumb in the millet fields of Liaoning Province near Manchuria, is largely ignored by paleoanthropologists who simply can't decipher the Chinese language and is also largely inaccessible, as yet, to those who would want to study it. Pope, who periodically visits the site, and *can* speak Chinese, says this fossil has been tentatively dated right at the time period during which modern man was supposed to be rising in sub-Saharan Africa: 100,000 to 280,000 years old.

Says Pope: "It's an unusual combination of features that nobody expected to see. It has a very large cranium—about 1,400 cubic centimeters—yet it has very pronounced browridges and extremely thin cranial bones. Usually you expect a prominent browridge to come with thick cranial bones. It also has what I called an occipital bun, a swelling in the back, and in the face; at the same time, it's extremely Asian, a very compressed maxilla, and very angular zygomatic bones meeting the maxilla at a right angle, which is distinctly Asian. So you have this combination of features—a delicate head sitting on the body of a bruiser—that no textbook would have predicted.

"It's absolutely unique, and it suggests that the out-of-Africa hypothesis is completely wrong. When I picked it up, expecting to feel a very heavy, thick cranium, instead it was like picking up paper—it was so thin and delicate I couldn't believe it. Jinniu Shan makes everyone go back to the drawing board. What

Jinniu Shan may indicate is that there is increased admixture, gene flow between East and West in the late Middle Pleistocene or the Late Pleistocene, which is completely different from wiping out a population and completely replacing it. The gene flow may have come as a result of increased migration across the Siberian plains."

Since China has more than the usual problems with dating, and this specimen was analyzed by the method of thermoluminescence, which many believe is less than reliable, there is likely to be heated debate about the Golden Ox Mountain Man. But its morphology alone is odd enough to warrant intense scrutiny. The second discovery causing excitement among Chinese workers is in the Shanxi Province, and known as the Xujiayao specimens, six to eight individuals, again older than 100,000 years, again distinctly Asian of face, found with stone balls that range from the size of a marble to that of a basketball—perhaps tools for processing hides.

"It seems highly unlikely," writes Pope, who ironically is friends with Cann, "that 'invading' populations of Homo sapiens would possess technologies or adaptive strategies superior to those of resident populations that had occupied a variety of Asian habitats for as much as one million years. This is especially true in Southeast Asia where forest habitats probably expanded rather than decreased in the Upper Pleistocene. Furthermore, it must be remembered that there is no dramatic change in archaeological assemblages throughout the entire span of the Paleolithic in the Far East. Nothing suggests the sudden introduction of 'superior' strategies."

Chris Stringer, who had an increasing investment in the replacement hypothesis, nonetheless describes the Chinese specimens as similar to transitional forms in northern Africa and "very important if they really are that old—more progressive than you'd expect."

That modern man may have risen in pockets across several continents as genes moved between and across populations, or that the human race moved as one entity in the same direction of modernity, but in fits and starts that varied according to region (causing some to acquire modern features while others went extinct or remained nearer to archaic form until later), both remain distinct possibilities. Archeologist Lewis Binford says he has seen no evidence of tool movement from Africa to Eurasia during the time forwarded by the Eve hypothesis. But whatever the case, Africa seems to have been a generator for certain of those new, modern genes—perhaps *the* generator, and in this sense perhaps indeed our birthplace. Despite the new finds in Asia, the African fossils from the period hovering around the crucial 100,000-year mark remain highly persuasive, displaying mosaic features—half-modern, half-archaic—like nowhere else but the Middle East for this apparently pivotal time period. And the modernistic features of archaic African specimens are closer to present-day morphology than those found nearly anywhere else on the planet.

The fossil record of Africa is such that it has led Mary Leakey to conclude that if nothing else, the basic premise of the Eve argument—a sub-Saharan origin 142,500 to 285,000 years ago—remains, "as far as the evidence we have now, a reasonable assumption." In fact, Mary and son Richard had both discovered fossils that lent great credence to the idea that a population of archaic Africans had evolved into the first anatomically modern humans before they had similarly evolved anywhere else. While the archaics in Europe had given way to Neandertals, the similarly heavy-featured archaics in Africa such as Ndutu and Rhodesian Man had made way for a more gracile form of premodern man, who, as I said, seemed more aligned to populations of today than virtually any fossils from Europe and Asia, and for the most part dated earlier.

Quite early in Richard's expedition during 1967 to the Omo River in Ethiopia, Leakey's team found two skulls known as Omo I and Omo II. These skulls were relatively recent, a throwback to the disputed, modern-looking Kanjera specimen his father had found and which may still prove to be significant if it can ever be dated properly. Right now Kanjera remains in a paleoanthropological limbo called the "suspense account." Anyway, those two skulls discovered many years later by his son in the Omo may one day be viewed as the most significant finds to have come from that particular locality in southern Ethiopia and among the more important of Richard's career, for they belonged to *Homo sapiens* and they appeared to be over 100,000 years old. They were nearly precisely in the range Cann, Stoneking, and Wilson would declare a full two decades later as the period during which modern *Homo sapiens* arose from a stock of African archaics. The extraordinary thing was that one of the skulls, Omo II, looked archaic, bearing some resemblances to Rhodesian Man and perhaps also to the old Indonesian specimens, while the second skull, Omo I, which had about the same cranial capacity (approximately 1,430 cubic centimeters), was almost completely modern and bore close relationships to the Cro-Magnoid representatives of the European Upper Paleolithic.

So at the Omo there you had it: an archaic next to something that was just about modern.

There was something breathtaking about the possibility that right there in similar Ethiopian deposits south of the Sahara were skulls demonstrating the transition from archaic *Homo sapiens* to modern ones. At the least, they hinted that archaics and moderns may have co-existed near each other. "I collected both specimens and they're exactly the same age," insists Leakey. "They're from the same strata on opposite sides of a river and you can see one site from the other site. I have probably collected as many fossils as anyone else around, and the one thing I do know is where things came from. And there's no question that Omo I and Omo II are the same age. The best estimate we could get from the geology and dating that was done in 1967, 1968, and 1969 would place them at between 100,000 to 130,000."

Whatever force it was that directed the Leakeys to the most significant sites also made sure that his mother would have her own fossil from this very same, critical era too. Mary's largely unpublicized discovery was in 1976, from that rich locale near Olduvai known as Laetoli—where the australopithecine footprints had been uncovered. It was a well-preserved cranium ("L.H. 18") that has been dated in the range of 90,000 to 150,000 years old.

Like Omo II, L.H. 18 was a very progressive type of archaic with a low, thickened vault and well-developed browridges. But the brows were only moderately thick and the glenoid cavities were fully modern. The Laetoli person was certainly less heavily built than Rhodesian Man, which investigators once described as Neandertal-like, and despite a robust appearance and some frontal flattening, it resembled recent humans in particular details of the parietal, occipital, and temporal anatomy. Though, upon studying a cast of this fossil in the Nairobi chambers, one is struck by some lingering *erectus*-like features, such as thickness of skull and perhaps some resemblances to Neandertals in the puffiness of the rear vault, the incomplete face has indicated modern characteristics.

Was Laetoli 18 another archaic turning into a modern African? "It's a bit of a conundrum," admits Mary. "It certainly has *Homo sapiens* features, but it wouldn't fit into a modern population. It's on the line towards *sapiens*. But it's not quite there. A lot of sites might belong to that time that are not dated. It's a very bad period in the record."

During the spring of 1989, the German paleoanthropologist Günter Bräuer from the Institute für Humanbiologie in Hamburg informed me that he and Richard Leakey were in the process of studying another exciting fossil from West Turkana that appeared to show the same transition from archaics to modern humans. In appearance it is more *Homo sapiens sapiens*-like than the Laetoli specimen. "I think it's very close to modern humans," says Bräuer. "And the dating is probably more than 100,000 years. One can see a lot! It's a partial cranium, a maxilla, it's a front, two parietal regions nearly complete, and the whole back part of the cranium, nearly the full occipital."

Bräuer has grouped Omo II, Laetoli 18, the new Turkana specimen, as well as another transitional cranium from Turkana known as "E.S.-11693" (for Eliye Springs) and a younger, modern-looking mandible from Dire-Dawa, Ethiopia—up near the Horn of Africa—all in a category distinguishing them as at the final stage of archaic morphology and therefore poised at the very threshold of anatomically modern status. Meanwhile from the Sudan has come a braincase that has been variously described as possessing modern and archaic traits but that, with a capacity of 1,500 cubic centimeters and other considerations to take into account, may already have taken the magical and nebulous step into modernity. Further north a series of fossils including one in Morocco called Djebel (or Jebel) Irhoud and another there known as Temara seem to have been transforming from the African clade of archaics into moderns.

The very breadth of transitional fossils in Africa and elsewhere has time and again called into question the idea that a small
isolated population could have served as modern humanity's sole
rootstock. Although there is a nice, catchy ring to such notions,
the situation was almost certainly more convoluted than what is
expressed by the Eve hypothesis, for while it is convenient to think
in terms of a single isolated population of these archaic-modern
types turning fully human and spreading outward—and while the
theory, spiced with good drama and high technology, may yet
prove correct—the simplest or most parsimonious conceptions are
often not the accurate ones. It could be forcefully argued, rather,
that the scenario was one in which modern genes cropped up all
over the continent, at first sparsely and nearly at random, the
pebbles scattered across a pond as divided tribes exchanged occasional genes, now and then creating humans who blended various
traits in unique new ways and interbred with the more archaic-
looking types who had not yet received the latest kind of DNA to
bring them better adaptations and closer to the modern threshold.
Fred Smith of Tennessee, who believes that in some ways the
transitional forms in Africa are no more impressive than transitional groups like the Vindija fossils in Europe, describes Florisbad, Laetoli, Omo II, and Irhoud as "a group which is in fact so
geographically dispersed as to make the idea of a localized speciation event leading to modern humans implausible."

But the European transitionals are younger than the African ones. Their kind are best represented by such specimens as the
Hahnöfersand remains in Germany, which are perhaps only
36,000 years old, and also by a postcranial fossil from Crete, which
might be over 50,000. Thus, with specimens like L.H. 18 at possibly 150,000, the Africans seem clearly to have gone through the
transition into modernhood far sooner.

Bräuer includes in his group a cranium with a relatively
high forehead disinterred from peats and coarse sand at Florisbad
in South Africa. If current indications bear out, this African archaic from Florisbad was turning modern 75,000 years before the
much heavier-skulled archaics in Europe gave way to the less-than-
modern Neandertals, let alone to any transitional specimens that
stood between Neandertals and moderns.

"Florisbad may be 200,000," says Richard Klein at the University of Chicago. "It's older than Omo and probably older than Laetoli. It is certainly older than 130,000, according to what we've found recently. It could turn out in the end that they're all the same age—130,000 or 200,000, but based on the information now, Florisbad is the oldest."

If Eve had been 200,000 years old, and if really there was a DNA Eve, such specimens as Florisbad, Laetoli, and Omo represented either her populations or the populations that immediately followed her. One of them, said Berkeley, was the population poised to turn fully modern and hug the Nile up to the Middle East before fanning in all directions to establish what we now know as the human race. There was the good chance, by this way of thinking, that Eve was Tanzanian, South African, or an Ethiopian. She may have stopped by Olduvai on one of her clan's journeys, nursing her children—our ancient grandparents—where all the *habilis* types—those first to widely use tools—and all the man-apes like *Zinj* had been! If she was 142,500 years old she could also have been a Kenyan, assuming the new Turkana specimens bear such an age out, or more specifically she could have been from the Mumba Rock Shelter in Tanzania, where very modernistic dental remains that are as old as Laetoli 18 have been found near an older archaic site called Eyasi.

"The archaic *H. sapiens* populations represented by the Florisbad, Ngaloba [another name for Laetoli 18], and Jebel Irhoud skulls probably evolved from yet earlier *H. sapiens* populations represented by skulls like those from Saldanha, Broken Hill, Bodo, and Lake Ndutu," according to Klein. "It is possible that one or more of the populations represented by Florisbad, Ngaloba, and Jebel Irhoud evolved into modern *H. sapiens sapiens*, perhaps through the population represented by the Omo skulls, and that this happened while Neandertals still occupied both Europe and the Near East."

These are not yet those promised and final few caves beckoning most loudly—they came next—but finally we were within a whisker of the first anatomically modern humans. "It is only in Africa," Chris Stringer has stated, "that fossils displaying a clear mosaic of non-modern and modern characteristics exist in the later Middle Pleistocene or early Late Pleistocene, and the presence of actual anatomically modern hominids in the early Late Pleistocene seems probable. Fossils such as Florisbad, Djebel Irhoud, and Omo II do seem to represent the kind of morphological precursors of modern humans that are missing from the late Middle Pleistocene or early Late Pleistocene of Europe. . . ."

Stringer, curator of human fossils at the British museum, emphasizes that while there was a lot of prehistoric action down there south of the desert, northern Africa cannot be taken lightly, for more than any other area, it "has fossils that bridge the gap between archaic and modern morphologies, and a secure chronological placement for the Irhoud and Dar-es-Soltan [also in Morocco] Aterian fossils would be invaluable in assessing the likely origin of modern human characters elsewhere around the Mediterranean."

But like those in Europe, the age of the crucial transitional forms in North Africa may be younger than those such as Omo further south. Moreover, instead of displaying features that are solely in line with those of other African archaics, some scholars have noticed Neandertal characteristics in the North African specimens, implying interbreeding between archaic Europeans and archaic Africans. Although many paleoanthropologists emphatically deny the existence of any Neandertal-like archaic forms in Morocco or for that matter anywhere else on that continent, Günter Bräuer believes that modern humans and Neandertals probably did interbreed north of the Sahara, making northern Africa and the Middle East start to seem like a melting pot. Bräuer sees Neandertaloid features in the mastoid process and in the occipital when observed from a lateral view present with the Jebel Irhoud fossil, which is believed to be 35,000 to 120,000 years old and which splendidly blends both archaic and modern features.

"Ten years ago nearly everyone thought there were Neandertals in northern Africa, but nowadays there's a trend to negate this," grumbles Bräuer.

As Bräuer imagined it, "During the Würm glaciation [65,000 to 95,000 years ago], anatomically modern humans expanded farther into the north—apparently moving out of East Africa. In both the Near East and northern Africa, there followed an intermediate period of mixing, after which these forms caused the disappearance of the Neandertaloid populations which had been living there. During the next millennia—in what was probably a relatively slow process of hybridization and replacement—they also superseded the European Neandertals."

Despite his strong feelings about interbreeding, Bräuer along with Stringer presented a cordial and tireless team of replacement philosophers who promoted the new genetic evidence with more vigor than any other paleoanthropologists. Stringer was an especially ardent booster of mitochondrial DNA, and his nervous chuckle at the thought of a mitochondrial mutation causing the Eve population to talk sprang from his understanding that one could suppose that the extent of genetic differences between one area and another area was related to their time depth only "assuming that most of the genetic data is neutral, assuming that selection has not played a big role in producing the genetic differences that we see."

He was going to hear from Wolpoff about this. Wolpoff was his chief rival in all of anthropology. It was true that Wolpoff had once rescued Stringer at a symposium in Yugoslavia when Stringer's argument that Petralona was not as old as some Greeks were saying caused one such Greek to "froth at the mouth," threatening the Englishman physically; but even if Wolpoff had bailed him out of that, their relationship had often been a bitter one. Even before the Berkeley data, Stringer and Wolpoff, set against each other simply by contrasting philosophies, had hurtled papers across the Atlantic at one another. It was at Stringer that Wolpoff had aimed his insult about Piltdown being another great discovery of the British museum. The exchanges were especially sharp in the journal *Science* (each warning the other to ignore their sets of data only at their own "peril"), and at one point when Stringer and his close English colleague Peter Andrews, also of the British museum, wrote a lengthy article for *Science* explaining the philosophies of multi-regional continuity versus their favored replacement model, Wolpoff led a contingent of eight scholars who gunned back with a long letter on August 12, 1988, complaining that Stringer had misrepresented their philosophy as well as the fossil record.

Whatever way that record was viewed, the fact remained that, just as many paleoanthropologists had long suspected, and just as geneticists like Doug Wallace were fully willing to weigh, current evidence continued to build that the initial thrust in the direction of modern physiology seemed to have come, if not from sub-Sahara Africa entirely, at least from somewhere south of the Mediterranean. Humans there were clearly on the brink of becoming *Homo sapiens sapiens* around 100,000 years ago.

And there was more genetic data in support of such roots. Using chimpanzees as an "outgroup" to better root the Eve tree, and comparing their nucleotide sequences to human groups, Thomas Kocher of the Wilson lab announced at the Asilomar conference in 1988 that DNA comparisons between that primate and humans confirmed that the deepest branches belonged to Africans. An African origin also seemed to hold up when another of Wilson's highly trained and capable workers, Linda Vigilant, finally did what the lab was criticized for not doing in the past: studying actual Africans. Vigilant, who is researching the genetic relationships between African populations like Pygmies in Central Africa and the !Kung hunter-gatherers from the Kalahari, says an analysis of sixty-one Africans (including some Nigerians and, in the case of the !Kung, including hair samples), conducted with the revolutionary new gene-amplification technique known as the polymerase chain reaction (along with a sequencing of their "D-Loops"), has shown the same general trends as the American black sample did. "Fortunately it's all consistent," she says.

But where Vigilant calculated the common human ancestor as living 238,000 years ago (and Cann was leaning toward a similar or even older number), Wilson now wanted a much younger age for the common ancestor. "The time we prefer is 140,000 for African Eve, not 238,000," he told me, contradicting his own lab worker. "Indeed, the time might be even shorter than 140,000. We are all derivative of one of the !Kung lineages."

Eve was from Botswana! As for the expansion of modern Africans out of Africa, an analysis of seventy Sardinians and forty people from the Middle East by yet another lab worker, Anna DiRienzo, indicated what appeared to be a swelling of populations in those regions between 37,000 to 52,000 years ago—suggesting, she assumed, an expansion of modern Africans in the Middle East and Mediterranean. Wilson preferred 35,000 to 75,000 years ago for the first major wave of Eve's migrating descendants. "We now have a more magnified view," added Wilson, referring to the new polymerase technique. And a finer calibration had refuted criticisms of his molecular clock, he claimed. The figures were bandied about by Wilson at a major meeting of geneticists on October 3, 1989, priming the pump for papers expected to be published by *Nature* and the *Proceedings of the National Academy of Sciences.* The news wires were hot once again, for the new evidence seemed to offer the irresistible picture of African females remaining firmly rooted until the major expansion, while the males had been wandering far and wide for thousands of years on what seemed like exploratory excursions. At any rate, Wilson, with a bit of modification and amplification, was standing behind his previous research, even if certain estimates tended to be slightly confusing, if not outright at variance, which led to some commotion in the lab. Meanwhile, other studies by other researchers could be used to at least partly buttress the Berkeley claims.

Although studies of DNA from the cell's actual center of heredity, the nucleus, have been stalled by its enormous number of genes and by the fact that recombination makes them difficult and murky evolutionary markers, the Oxford group led by James Wainscoat has already published an analysis of the type of genes, beta globins, responsible partly for hemoglobin production. Wainscoat's group, looking for "haplotypes," or sets of mutations that characterize certain populations, has observed that Africans had a certain haplotype not found in any other population, and in looking at other haplotypes, along with genetic distances based on changing gene fragments, or polymorphisms, suggests that there had been, as Cavalli-Sforza now agrees, a major division of populations into an African and a Eurasian group many, many centuries ago.

"We're doing another nuclear DNA set of polymorphisms which show more variation and so give more data," says Wainscoat. "And we've done more populations. We find exactly the same overall conclusions, that Africans are extremely different from Europe and America, and the other thing we find is that there's more intrinsic variation within an African population. We were careful to say in our original paper that the fact that there is this split between Africa and the rest of the world is consistent with the idea that man originated in Africa, although it doesn't actually constitute any proof."

Because there is a chromosome in the nucleus, the "Y" chromosome, which, as an opposite of mitochondrial DNA, contains genetic material passed along to offspring only by the father, there are those like Cann who suggest that the common ancestral man may one day to tracked just as Eve supposedly was. The search for Adam. Wainscoat had alluded to such a possibility in his review of the Cann, et al., paper in *Nature*. "Several people have actually taken it seriously and are working on the concept of 'Adam,' and I just made it as a joke," says Wainscoat. "One group wrote me and told me they didn't think these two [Adam and Eve] could have met!"

But according to workers at the Laboratoire de Génétique Moléculaire in Paris, research on the Y chromosome already has shown distinct areas unaffected by recombination, which would lend itself to evolutionary analysis, just as mitochondrial DNA does. There is no doubt that, with the massive federal project under way in the United States to map the entire nuclear genome, spearheaded by James Watson, the very man who'd discovered DNA's structure, a windfall of information about these clandestine compartments of heredity and their association with man's development will soon be forthcoming. The polymerase technique (far faster than old restriction mapping) will cause a cascade of new information. In the meantime another nuclear investigator, Jeffrey C. Long of the University of New Mexico, trained in both genetics and anthropology, bolsters the idea of an African origin. Long and four associates have evaluated the literature and presented it with their own laboratory typing on 222 chromosomes culled from Africa, China, Greece, India, and Thailand. They are ready to report that seemingly ancestral haplotypes are missing from Asian samples and present in African ones.

But there were those other geneticists besides Wallace who implied an origin in Eurasia, and it was time for us to leave the DNA labs and head for the caves ourselves. Right now there were really only the fossils to look at. We had to rely on the unreliable. So far Eve wasn't to be found in the DNA. She just wasn't there. In the wake of questions over the transitional forms from Asia and Europe that seemed to be turning modern like in Africa, in the wake of genetic evaluations that were at times completely at odds with each other, in the wake of the growing feeling that mitochondrial DNA was not the magic bullet it seemed at first, and in the wake of questions about Wilson's assumptions and the molecular "clock" itself, it seemed that the data was saying as much about the limitations of current genetic research as it was of evolution itself, and that therefore it was wisest to place the Eve hypothesis in what Louis Leakey, when unsure of a fossil, called the "suspense account."

But that wasn't to say there was no Eve. And while Africa was still the strongest candidate (in most people's minds), there was the chance—always subject to sudden and drastic change— that Eve had been where many people always thought she had been: in the Middle East, land of the Bible and that chapter called Genesis.

That's where the beckoning cave was: Israel.

20

I<small>T IS UP A MOUNTAIN</small> just south of Nazareth looking over the kibbutzes of Emaq Yizreel. The boulders are heaped and steep. The gnarled brush hides scorpions. Small trees are bent as if by the sun. There are vipers, too, and not far away are the troubled villages of the West Bank. But here there is total quiet, even desolation. It's a grueling climb. There are few bushes for support. The knees ache. The sun is merciless. Dust rises from a quarry that is now far below. There is a small ravine to hike across and a lot of precarious dolomite.

But finally it is visible, looming without mistake above a gravel fall. The cave is a gaping, oval orifice cut into the imposing, crumbling cliff face, and though a new discovery somewhere else would diminish its magic, it is increasingly likely that this time-worn outpost overlooking Israel's northern valley was that long-sought threshold—the place where certain of the last archaics turned into the first *Homo sapiens sapiens.*

It was also Eve's final resting place. The French call it *la grotte de Qafzeh* (pronounced "Kaf-sa"), which comes from a most appropriate Arabic term that means "mountain of the jump." It was the jump into modern anatomy. There had been a cemetery at the cave's terrace, and that's where the remains of fifteen or so people were excavated, mostly in pieces. Of all the ancient bones removed, the best belonged, so very appropriately, to a female.

She was only a teenager when she met her end, probably

115,000 years ago. Her grave was exhumed by a team of French researchers in the 1960s, and two decades later, in 1988, the deposits were found to be much older than previously thought. In 1989, new tests showed them to be more ancient still, making her people the oldest anatomically modern population known to the world. It seemed like Eden just might have been where the Bible said it was.

After an exhausting climb up that stack of dolomite, sucking in the dry air of Eden, I went to the Rockefeller Museum across the road from the wall of Old Jerusalem and had Eve's skull unlocked from a bin.

"Do you want to see her mandible, too?" asked Joe Zias, the curator.

Eve's mandible! "Yeah," I fairly panted.

Zias left with his keys, was gone for a few excruciating minutes, and came back to a workbench with the skull. He was nonchalant about it. I'd expected him to kneel reverently as he presented Her. There in front of me, to touch and wonder at, was "Q.H. 9," that grandmother of everyone, plucked from the terrace of the cave, the first real woman.

Or at least she had my vote. The setting was too authentic for it not to be her. There were the archeologists huddled over boxes of relics and skeletal material, there were the museum's classic corridors with bleached archways, and there were the signs everywhere that said "No Admittance," just as a sign at the Garden of Eden would now have prohibited admittance. Blocks and busts and stones.

But the skull was far more dramatic than anything else, dark from fire or iron oxide—a blackish rust color. Otherwise you couldn't tell it from a medical school skull. If her face gave just a hint of jutting out, that was one of the few remnants of an archaic past, for she clearly had a rounded chin, the mandible was quite like ours, her cranium was beautifully smoothed over—nothing like a craggy *erectus*—and to my relief she had little in the way of a browridge.

How much more refined was her face than the thick broad cheeks of a Neandertal. I wondered at the integrity of her teeth. There was the hint of a smile. After observing so many fossils that were mere pieces of crushed maxilla, it was remarkable to see a specimen with its teeth intact. They were all there. They were pretty darn straight, and I didn't see cavities. "Well, you know, she was only eighteen," explained Zias, the curator. "And back then there was no Wonder bread or Velveeta cheese."

Eighteen: it seemed so young for our 5,750th great-grand-mother.

Eve just stared vacantly, humble and elegant, delicate, oblivious to the chatter of archeologists on a tea break down the hall. Perhaps she was reminiscing about the old days when she spent her days scraping hides and scanning the horizon from what was probably a temporary stopover nestled between the hills of Galilee and Mount Carmel. From up there the view had been as expansive as it was beautiful. Israel back then was geologically much as it is today, but instead of corn and cotton in the lowlands, instead of the kibbutz, there had been great herds of horses, deer, and gazelle, some elephants too, a lot of sheep, a good number of nagging goats and grunting boars, perhaps an occasional rhino.

It was the animals that have led to the dates. First it was mice, but most recently it has been those deer and gazelle. According to Henry Schwarcz, chairman of geology at McMaster University in Canada, the teeth of a deer or other mammals left to fossilize in a cave can function as a timepiece in that geologists can measure how the enamel has been affected by radiation in the soil. This is known as "electron spin resonance." The signal coming from an ancient tooth grows in size the longer enamel has been exposed to background radiation. "It's not the highest precision, but nevertheless we have quite firm an estimate, and the date is 115,000," Schwarcz assures me, "plus or minus 15,000."

Which meant she might be as old as 130,000. If the dating holds up, that would surpass her close competitors from Africa. Although the ranges of time for Qafzeh overlap with fossils found way down on the Cape of South Africa, at Klasies Rivermouth, those fossils are very fragmentary and therefore subject to more doubt. Another candidate is Border Cave, which is in the Natal Province. But Border Cave appears to be younger than Qafzeh. According to Karl Butzer, a geologist from the University of Texas in Austin, Border Cave is 65,000 to 95,000 years old, which is now below the lowest range for Qafzeh. In East Africa, Omo I and scattered fragments such as Mumba Rock Shelter suggest that modern genes may have originated there, too.

But considering its size, which is one forty-fifth that of Tanzania, Israel presents a more dramatic scenario. In that sliver of a country betwixt Egypt and Lebanon—the only landbridge from Africa to Eurasia—is what archeologist Arthur J. Jelinek thinks might be "the greatest diversity of Later Pleistocene hominid fossils in a limited area." If Israel didn't have man-apes and *Homo habilis*—well, there was a trend afoot to look at those more as ape than human anyhow. Who, again, wanted an ancestor that still slept in trees? Lucy was closer to a chimp than she was even to the oldest specimens from Java. It was *Homo erectus* and archaics that always drew us back, and Israel had what appeared to be one or two *erectus* teeth and maybe a bit of cranium at Ubeidiya near Kibbutz Afikim not far from the Sea of Galilee that were probably older, at up to 1.4 million, than Peking Man. A date of 1.4 million, in fact, wasn't much younger than the Nariokotome Boy. And even if it wasn't quite old enough for anyone to claim that *erectus* had originated in the Middle East, it was not the origin of *erectus* we were finally concerned with as much as the progression of *erectus* descendants into *Homo sapiens sapiens*.

And here Israel hit pay dirt.

It was nearly an embarrassment of riches, the Turkana of the Middle East. Israel had specimens that not only bridged gaps in time but also seemed to bridge some gaps in space: between Europe, Asia, and Africa. There were archaic remains here that bore resemblances to the pre-Neandertal archaics of Europe— Steinheim, Arago, Ehringsdorf—while Qafzeh looked more like present-day Africans than did African fossils from a place like Border Cave, in the view of Erik Trinkaus. According to Chris Stringer, Qafzeh also had traits that put him in mind of the remote Australian specimens. Was this sliver of contested turf, this land that was holy to Jews, Moslems, and Christians alike, this territory that threatened the start of World War III because of an incredible attachment people felt for the place, was it the crossroads and meeting place of human genes just as it later served as the crossroads and place of exchange for religions, agriculture, and cultures?

Or was it the catalyst itself? Was there not the chance that the Levant had integrated genes from all inhabited parts of the world and had come up with new seeds that it then planted to the west in roomy Africa, where the sun sprouted them like eager spring shoots, and to the north in Europe and Asia?

Odds were perhaps that it wasn't the main origin, but Stringer himself, a proponent of an African birthplace, said in a recent paper that if the Qafzeh material proves to be older than 50,000 years—as now appears to be the case—"the Asian corridor linking the three continents of the Old World might also become a critical area for the *origin* of modern humans, rather than for their dispersal."

Somehow, out of the confusion of DNA and paleoanthropology, there was the unsettling feeling that everyone had rather ignored this place, even though it is the perfect compromise between Asia and Africa. It was not hard to imagine why it's ignored. Israel isn't the trendy place of origin, and scientists as much as anyone else are creatures of trendiness.

But the Middle East deserved more than passing notice, and at the very least it had served for more than a million years as a membrane between Africa and the rest of the world. If *erectus* had come from Africa, it was certainly more probable than any other route to have transversed the Near East on the way to Java. There were no Boeings in those days. And if the geneticists had trouble deciding whether Asia was separate from Europe and Africa, or whether it was the other way around, one constancy remained intact: there between Africa and Eurasia was the landbridge known as the Levant.

"This is a meeting place, this is the Levantine corridor which connects Africa and Eurasia—the place where faunas were passing in both directions beginning with the Miocene," says Harvard archeologist Ofer Bar-Yosef. "If you go from Africa to Asia this is one easy way to cross, because otherwise you have to cross in the area of Yemen, and then you still have some deserted patches to go. So humans in early stages are following whatever subsistence patterns they had, whether hunter or gatherer, and they could always find a refuge in the Levant. This is an incredible place. It has more plant life and bird species than California."

According to Bar-Yosef, who considers the Levant "an absolutely viable option" as the birthplace of modern man, the plant life shows just what kind of fertile meeting place Israel is. There are 2,530 species of plants in Israel, he points out, or 0.085 species per square kilometer, where a place like coastal California has less than half that number (0.036) and Italy less than a quarter. Others cite the camels that migrated to Africa all the way from America, as well as deer that apparently came from Asia. According to Geoffrey Pope, "there's no question the Middle East has been a place of major gene flow for a long time. And it's true of the animals. If you look at Ethiopian rodents, what we find is that a high percentage are Asian—that they've come in from Asia."

Bar-Yosef believes there may be an analogy between the movements of ancient humans and the movements not only of animals but also, later on, of the one factor that separates us from the uncivilized: ideas. "No one would like to see it coming from the Near East, but I worked also on the origins of agriculture in the early Neolithic," says the Harvard scholar. "The origins of agriculture took place in the Levant due to certain circumstances which combined topography, climate, degree of foliability, ecological considerations, as well as social considerations, and the origin of agriculture took place between Damascus, Jericho, and south of Jordan. From there it distributed from the Levant into Turkey, then into Europe, and from the other side into Pakistan. Why not use the same model for the origin of modern humans? This is a meeting place, where maybe the whole shtick happened. But this would mean Israel would again be at the head of news, and people don't like that."

Certain paleoanthropologists still wanted Europe to be the center. Many more preferred Africa. A minority were rooting for Asia. And once such a view was set into place, many of these scientists spent their careers molding interpretations of evidence to fit those views—and unabashedly doing so with healthy doses of intuition.

But whatever they did or did not prefer, there was no denying that Israel and other parts of the Levant had some wholly

baffling fossils going for it. Its human remains combined a constel-
lation of archaic and sapient features in such a striking way—
integrating characteristics from all three continents—that even if
Qafzeh is one day soon proven to be younger than newly
unearthed specimens from somewhere else, the mosaic of early and
modern features will continue to maintain Israel's often unmen-
tioned stature as a hotbed of paleoanthropology. There are, for
example, the thigh bones from Gesher Benot Ya'acov that seem
to represent late *erectus* or early archaic, dating to perhaps 600,000
B.C. At a site called Zuttiyeh about a dozen kilometers from the
teeth at Ubeidiya a fragmentary skull was found, indicating the
existence too of broadly ancestral archaics that roamed the Middle
East 150,000 years ago.

So we have a continuum. And it reaches fever pitch at
Mount Carmel. In this range of splendiferous vistas (which served
as a refuge to the prophet Elijah) are four fossil sites that were
stratified with tools, ash, animal bones, and human remains in
such a way as to present, like pages in a rarefied book, the physical
and technological changes over the course of at least 120,000
years. One of them, Mugharet et-Tabun, may reach as far back as
the 2,200th century B.C., according to Arthur Jelinek.

These and the other Mount Carmel caves, in many cases
within a short walking distance of each other, grant us a succession
of tools that stretch across the archeological board, from the
Lower to Middle to Upper Paleolithic. Start with those old
Acheulian hand axes most frequently associated with late *erectus*
or subsequent archaics and work your way through dizzying com-
binations of deposits containing industries that include a Mousterian phase resembling that of the European Neandertal and an
"Amudian" phase with those blades and other artifacts which
have been linked to the emergence of modern man (the Cro-
Magnon) in western Europe.

So too can a succession be argued for the fossilized humans. A mere 70 meters from Tabun, chockful of more tools, is
a huge cave called Mugharet el-Wad, and just 100 meters east of
that cave is a famous site called Mugharet es-Skhul, where, during
the 1930s, ten individuals were exhumed from another ancient
cemetery. The oldest may date at least to 80,000 or 100,000 B.C.—

long before Elijah. "This rock shelter is very important because it shows a clear sapiensation of Mousterian people, they have real modern human features, like the chin, like the length of the long bones, like the reduction of frontal torus," says Baruch Arensburg of Tel Aviv University, pointing to a spot just outside the cave.

Indeed, one of the original excavators, T. D. McCown, observed that if only teeth and skull anteriors had been found at Skhul, the individuals might have been declared Neandertal. On the other hand, if only hands and *feet* had been unearthed, they may have been declared fully modern, like the Cro-Magnon that came immediately afterwards. To further appreciate just what a mix this was, consider that the remains at older Tabun shared not only certain physical characteristics with those at Skhul, indicating intermediate forms, but also certain of the same cultures.

Some looked so Neandertaloid that it seemed like European cavemen had migrated south to the Levant during periods that got too cold—wearing their French summer shorts, joked Arensburg. But if large browridges were still to be noted in one Skhul specimen (Number 4), it had no characteristic Neandertal bun at the rear of the skull nor much in the way of the trademark jutting Neandertal face. And the Skhul population (at least in the case of Number 5) was getting higher of forehead and very much like what was to be found at Qafzeh, where there were also specimens that mixed in archaic traits but where brain capacity was heading toward 1,600 cubic centimeters and where Q.H. 9, a.k.a. Eve, had already achieved modernity, according to many enamored morphologists.

The transition to anatomically modern *Homo sapiens sapiens* may also have occurred at N'ame (or Na'ame), Lebanon, near Beirut, where rumor had it—and rumor only, since the date was highly suspect—that a site of the same approximate age as Qafzeh contained identical implements, namely, the same short Levallois points. At Skhul there were also weapons. A wound to the left hip joint of one Skhul man was thought by McCown to be "unequivocal testimony to their possession of spear-like weapons of great penetrating power." At Qafzeh, meanwhile, an eleven-year-old had been nursed to health after suffering a piercing, life-threatening wound to the head. Taken as a whole, along with remains found in another Mount Carmel cave, Mugharet el-Kebara, as well as with remains found in other parts near the Mediterranean from the Irhoud and Temara fossils in Morocco to remains sweeping east into Iraq and up to the Crimea and Afghanistan, one forms quickly the picture of a great but varied transition, the skeleton becoming more slender, the face toning swiftly down as the transition is made to modernity, the ribs getting thinner, the thumbs less powerful, the people caring for their elderly and infirm as they increased in brainpower.

Assuming the dates are not skewed, the great mystery is why modern specimens like Eve seem to have existed *before* some of the more archaic humans. Indeed, her population was immersed not in technologies normally associated with modern cavemen but rather into the Mousterian culture, which, again, has mostly been tied to Neandertals. And the perplexing 115,000-year date places her before there even were any classic Neandertals in Europe. Did she leave and let some leftover archaics take over at Neandertaloid places like Kebara, which has been dated at only 60,000, or did they co-exist—moderns and archaics—for many millennia, with occasional interbreeding?

"Sometimes there was mixing, sometimes there was developing alone, and the result of 100,000 years of this development is that today we find some groups like Qafzeh and Skhul that have *Homo sapiens sapiens* features and some groups from the same forefathers who have Neandertal-like features," says Arensburg, who teaches anatomy at Tel Aviv University's medical school. "In my opinion this is one demonstration that different levels of development from *Homo erectus* to *Homo sapiens sapiens* were taking place in the Middle East, and one of these levels was the so-called Neandertals from the Levant. I can see more developed tribes of *Homo sapiens sapiens* and tribes that were more Neandertal or *erectus*-like roaming around, and if you take a more Neandertal woman, why not?—there was occasional interbreeding."

Continues Arensburg: "At Kebara there was no flint, so they had to go ten or fifteen kilometers to get it for their tools, and on the way they may have met another group, and maybe they traded animal skins and women. There is no doubt they were communicating—no doubt. The Mousterian man from Kebara, in

Unit One, gave us a special opportunity to prove if Neandertals or Mousterians were or were not talking, and we found for the first time in the history of anthropology a bone at the back of the tongue indicating that the speech organs of Mousterian men were absolutely the same as the present day. Tools from the Negev to Galilee show the whole country was inhabited by small groups. There was a diffusion of culture and with a diffusion of culture was probably also a diffusion of genes."

There was the image of gracile Eve in the heights of Qafzeh looking to the far, swampy horizon as a troop of less modern beetle-brows strutted clubs in hand. Matted hides were slung over their shoulders. They carried sharpened sticks and prismatic blades. As they searched for a good deposit of flint or a new spring or a cave with a better view, they were also inching their way toward the Balkans or in the other direction to Egypt and Morocco.

It's a nice, simple image, but there are questions about the age of Q.H. 9 (some think she may have been a girl of only fourteen), questions about her robustness (she may have needed quite a touch of makeup), and questions even about whether "she" was really a female! Though our Israeli Eve gets the most votes as the best current specimen of the oldest *Homo sapiens sapiens*, there still is not unanimous agreement over whether Q.H. 9 is fully modern. One reason is a fairly incredible one: paleoanthropologists have not yet been able to devise a comprehensive and universally acceptable code for what constitutes a modern human.

Were we going to end up, after all this, back in sub-Saharan Africa?

"I excavated many, many skulls, and many skeletals, and I'll tell you the truth," paleoanthropologist Yoel Rak assured me, "if Qafzeh 9 was excavated in a modern graveyard, you hardly would pay any attention to it."

But the same could be said of certain African specimens. There were two places that vied with Qafzeh down there, and one of them was Border Cave. Situated just below a crest of the Lebombo Range above the Swaziland lowveld, the site, first investigated by Raymond Dart in 1934, has yielded four fragmentary individuals of interest to us, for their crania and mandibles are unequivocally ascribed as modern *sapiens,* and if a more recent date is now assigned them, there are those who in the past have pegged their age at right around the time now granted to Qafzeh: 110,000 years, give or take 10,000. A number of morphological similarities to modern indigenous populations exist, according to South African scholars, "and these suggest that, in view of the present dating evidence, the fossil series may be regarded as a still largely undifferentiated basal stock, from which both the South African Negro and Khoisan peoples, amongst others, ultimately arose."

Oh, to spoil all the fun: isolated teeth from two other South African locales, Die Kelders and Equus Cave, are perhaps also from this time period and are also referred to as belonging to modern men. And the mandible and skull fragments from Klasies Rivermouth, about 600 miles southwest of Border Cave, at the tip of the cape, are now "dated fairly secure on geological correlation" at 115,000 years, says Richard Klein.

Others like Lewis Binford say they are only 80,000, and still others say there are resemblances (in specimen "41658") to Skhul.

Peered at with a microscope, one of the fragments reveals what appear to be parallel cut marks inflicted by stone tools and suggesting that the face was defleshed by a primitive blade, indicating possible cannibalism, or that the scalp was taken in ritual. There were all kinds of neat things to be found down at Klasies. The tool artifacts, though including more blades, resembled the Mousterian stage in Europe (again the one associated with Neandertals, even though the Klasies people were decidedly not Neandertal, in the view of most), and subsequent tools included crescents and trapezes that may have served as barbs for wooden spears. Certain of the artifacts, according to British archeologist John Wymer and anatomist Ronald Singer of the University of Chicago, "show that such cultural development was evident in this region several tens of thousands of years before it appeared in Europe."

Others argue that the blades are different than what was found from subsequent periods in Europe—fashioned by a different technique. "The people from Klasies were not modern in their behavior," stresses Klein. "That's why they were confined to Africa for so long."

Although, unlike Skhul and Qafzeh, where in some instances people had been buried with their arms folded and knees bent, along with trophies or amulets like a boar's jaw, there were apparently no cemeteries at Klasies itself. But was there ever a lot of leftover food! The Eve here was virtually a mermaid. While the Klasies people weren't actual fishermen, marine birds were caught on the beach and these prehistoric connoisseurs liked limpets, brown mussels, Turbo, small periwinkles, and black-backed gulls. Mainly the limpets and mussels.

"The Klasies people clearly exploited the marine environment and they did so successfully at a very early date," says Philip Rightmire. "In fact, Klasies provides the earliest evidence for an exploitation of marine resources. The people were taking shellfish of a number of varieties. They also seem to have taken flightless birds—penguins, for example. And sea mammals such as the seal, things like that. There is evidence for burning at Klasies, and a number of bones are actually charred or blackened. There are no implements that suggest sewing. Nothing suggests the use of arrowheads. The Middle Stone Age tool assemblages are rough and ready. Presumably the Klasies people had access to skins and hides and used them for covering. Probably it was effective to hunt in small parties. It would take several people, for example, to drive a group of eland or a small herd of antelope into a blind or over a cliff. That might have been one of the more effective ways of taking larger mammals—simply scaring or driving the herd over a cliff into a gorge. And that," adds Rightmire, "would take cooperation on the part of the people."

Maybe the women collected the shellfish. Maybe there were man-made shelters, like at Terra Amata, but so far there is no evidence for them. There's no artwork yet. And were the Klasies people really modern? If one took a hammer to Laetoli 18, the pieces might look just as modern, argues Wolpoff, a worry that also concerns Klein. "All it's telling you, the mandible, which is the principal fossil cited there, is that the people had relatively short flat faces, particularly compared to Neandertals," says Klein. "The mandible is modern but we don't know what the rest of the skull and body looked like. I think the North African evidence like Dar-es-Soltan, which is associated with artifacts over 80,000 years old, is more convincing for the origin of the anatomically modern human."

Whatever the final academic fate of Klasies—and the final dates for Qafzeh and Border Cave—they stand as indications that modern man in the Middle East and Africa was sending modern genes into a worldwide pipeline, and so finally the Israeli and African specimens lend us the best notion of how man became man. Perhaps the genetic static crystallized in one specific spot into modern humans, crossing the threshold into *Homo sapiens sapiens* before anywhere else, but in the end that was basically unknowable—what was the definition of "modern man"?—and finding Eden, with current technology, was like looking for the limits of the universe with a pair of Sears binoculars.

Though without doubt there were cases where groups did supplant other ones, and modern bands which expanded outwardly into something akin to huge, loosely connected tribes, the rise of modern man probably occurred as Karl Butzer suggests: a genetic transfer between marginal and overlapping populations that sent new genes from one area to another speeding up the tempo of "evolution" without any massive movement of people or any worldwide expansion by a single tribe. Modern humans could crop up all over a continent, now and then creating humans who blended various traits in unique ways and interbred with the more archaic-looking types who had not yet received the latest kind of DNA to bring them closer to the modern threshold.

In the long run, no connected continents developed in a vacuum. It was a two-way street, and from most appearances, it was a rather continuous one. If Africa was responsible for distributing any number of new traits, surely it also accepted some. No matter the distance, the populations were connected as if by an invisible thread. Hints of the rear skull puffiness copyrighted by the European Neandertals were also present in specimens as diverse as Africa's Eliye Springs, Laetoli 18, and China's Jinniu Shan. Didn't Petralona from Greece bear resemblances to Bodo down in Ethiopia and Broken Hill in Zambia? And what about Narmada in India, which called into mind not only Broken Hill and Petralona, bring-

ing together Europe with Africa, but also that vexing Dali specimen way out, again, in China?

If evolution as portrayed now for more than a century is
correct (and the turmoil within evolutionary studies, especially
with dating, was enough to spike anyone with nagging doubts),
somehow the genes with modern features carried an advantage
that allowed them to predominate. Perhaps these genes were ping-
ing among the populations like marbles in a pinball, touching here
an Asian *erectus* to develop a lighter skull, a Neandertal at Hahnöfersand to lose his browridge, a Tanzanian archaic to grow a higher,
smoother forehead.

So good were these new genes, it seems, that modern man
was allowed to expand at a previously unseen rate, migrating or
simply settling down to create the nations and races, such that by
40,000 or at least 12,000 years ago man was a differentiated but
fairly uniform species that extended now from southern Africa to
Australia and the Americas. According to the Wallace lab, the
Americas were probably settled by just a handful of extended
families that crossed the Bering Strait when the ocean was low
enough during glaciation to form a bridge of land. "What we've
done," says Wallace, "is study Indians from North, Central, and
South America and found that they share a number of rare Asian
DNA types and have some unique polymorphisms that are found
only in the Amerindians. So what we're basically finding is that
there appear to be three major maternal lineages that are giving
rise to all the different tribes we've looked at so far."

As Indians were settling the Americas, there were the Cro-
Magnons adding last flourishes to the cave art in Europe, iron
pyrite showing up in Belgium, clay pots in Japan, and Stone Age
humans developing permanent communities even before the advent of agriculture. The domestication of animals and systematic
growth of barley and wheat, which began in the Middle East about
10,000 B.C., brought many more people together in ever-expand-
ing, increasingly complex social entities that were no longer shack-
led to the hunter-gatherer lifestyle, and now were centralized to the
point of establishing large villages and later cities. The more people, the more ideas. That was the real definition of modern man:
ideas. It was the ability to invent more than just a hand ax that

established *Homo sapiens sapiens* as the most dominant species in the history of earth, and with farms and permanent homes to protect, the occasional caveman skirmish gave way to a disheartening and unending record of war.

When races developed is anybody's guess, but geneticists have come up with interesting indications of how closely certain peoples are to each other. According to Masatoshi Nei, such European populations as the English, Germans, and Italians are very closely related to one another. They are more closely related, in fact, than Chinese are to Japanese. Meanwhile the English are also closer to people in northern India than Eskimos, for example, are to Alaskan Indians. Saudi Arabians are rather distinct from everyone, and so are Finns, while the Basques in Spain are closer to neighboring populations than their language and looks imply; South American Indians show a perplexing closeness to Polynesians; the Polynesians in their turn are genetically similar to Chinese and Indonesians; Papuans and aboriginal Australians form a cluster but with significant genetic differences between them; American Indians are more distant from Mongoloids than generally thought (perhaps because of inbreeding); and North Africans in Libya and Egypt are more closely related to Europeans than to sub-Saharan Africans.

Others have found that despite being scattered around the world for two millennia, Jews remain genetically distinctive, closer to each other even when separated by thousands of miles than they are to non-Jews among whom they have lived for centuries.

Overall, Nei and Gregory Livshits of Tel Aviv University reported in 1988 that the smallest genetic difference between major groups is between Europeans and Asians, and the largest difference is between Asians and Africans.

But the statistical methodology in studies such as that of Jews, as well as the racial undertones, make such matters subject to wariness and dispute. And they obscure the larger and far more important finding, agreed upon by a number of laboratories, that the differences between races are not much more than differences between members of the same population, indicating a huge mosaic of brotherhood. Nei himself concludes that "the genes in the three major ethnic groups of man are remarkably similar, although the phenotypic differences in such characters as pigmentation and facial structure are conspicuous." In other words, if there was something genetics had told us about human development, it was that we're brothers no matter how anyone looks.

I still had my mind on Eve. And I wasn't about to forget East Africa. That land where Lucy hid in trees, where robust aus-tralopithecines munched with those nutcracker molars, where *habilis* supposedly made the first flakes and choppers, where the Nariokotome Boy got carried off by a flood, where Laetoli 18 set a course toward modern looks, improved upon by Omo I and probably the population at Mumba Rock Shelter in Tanzania: I had to make sure she wasn't there.

And so from Israel I had boarded El Al to Nairobi, had that tea with Mary Leakey, searched through the anteroom of the national museum with the help of her son Richard and his wife Meave. I handled the 17-million-year-old skull of the ape *Proconsul,* stared dumbly at "1470," sifted through any number of other fossils including a cast of Laetoli 18, which simply looked too robust and low of skull to be Eve. Alas, she wasn't anywhere in the hundreds of specimens locked carefully in cabinets that lined the padded fossil room. Leakey himself warned me of that. "I think there's a great danger in using a single example as a basis for saying this is the earliest *sapiens sapiens,*" he told me in his spare, modest office, which looked like it belonged to a department head at a small university. "There is an impression developing that *sapiens sapiens,* or the modern expression of *sapiens,* is present in various parts of the world at about the same time."

When I told him about a rumor I'd heard that a team headed by his rival, Donald Johanson, had been working in Tan-zania and had unearthed a modern man from the critical 120,000-year period, which would have been just what I'd come for, Leakey only smiled a faint smile, refusing still to offer encouragement. "Well, they're very secretive, the group at Olduvai now," he sighed. "But yes, I've heard they've found a skeleton, and there's a Tanzanian who's describing it and who thinks it's just an intru-sive burial. He says it's absolutely modern and he thinks it's only a few thousand years old."

I should have known. I had already been out that way. I had hopped in a cab for the border of Tanzania, past the misty highlands of Kilimanjaro, hiring a Land-Rover and a Swahili guide for a tortuous journey over roads that had not been repaired since 1962, mere patches of incredibly broken and missing asphalt, having to go off into the savannah when the road ended or became too bad, taking a day and a half of non-stop bumping and grinding to go 200 miles. The only civilization had been in Masai huts, which were made with dung and mud.

There were groups of twenty or so natives in the tiny settlements, just like in olden times, and warriors in red capes approached with spears trying to sell us ostrich eggs.

I thought: Eve must have used ostrich eggs for something.

And the Masai women, in sackcloth and doing a lot of heavy lifting, carrying huge loads of firewood at the same time they tended straggling offspring: how tired Eve must have been, how absolutely pooped, that star-laden night so long ago!

I had passed galloping giraffes and the spindly gazelle Eve must have butchered, and I stopped amid the soda mist of the Ngorongoro Crater to get a close view of hippos in a water hole.

Eve must have come across hippos taking a bath!

There were elephants standing nearby among the tall swamp grass, mute to my mission, moaning for their survival, vultures overhead, hyenas always residents of the upper veldt, vervet monkeys chittering and spilling from the branches, just like in Eve's day, leopards somewhere in the surrounding jungle, a rusted Land-Rover rotting off the side of one cliff near a memorial to conservationists slain by poachers or rhinos, dust devils on the thirsting horizon, like little tornadoes—sky larger than anywhere in the world, bluer than anywhere in the world, with birds as colorful as anywhere in the world, into a territory where there weren't even many Masai but instead just a barren terrain of withered scrub and old thorn trees that whistled with the dry wind near the skeleton of a giraffe that had been devoured by lions.

Then across another stretch of dust-belching savannah, there was Olduvai Gorge. Layers of volcanic cake. Stone scrapers underfoot. A plinth and marble slab where *Zinj* was discovered.

I looked toward the distant highlands beyond which Laetoli and those footprints were hidden, where the termite nests were like huge sandcastles, where the grass was taller but no less haggling, where the land was hillier, the air cooler, the fossils older—and called out for Eve in a shrill psychic voice.

She wasn't there either.

And that evening, under the dark blue and starlight, amidst unbridled heavens and uncounted pasts, there was a visceral knowledge that went beyond logic and classification, belonging instead to the faculty of intuition, which barked from deep within: no one knows and perhaps no one ever will, for the past is prologue and the past is past.

NOTES

CHAPTER 1

For an excellent description of how man evolved, see Bernard G. Campbell's *Humankind Emerging,* 5th edn. (Boston: Scott, Foresman, 1988). I saw some of my vague fantasies materialize in the form of better-put and actual fact in this exceptional textbook, which was especially useful on the tactics of ancient hunters. Naturally the scenario comes from dozens of other paleoanthropologists too. Also, I visited the outback territories of Tanzania.

CHAPTER 2

The DNA theory, as we see in the next chapter, is in *Nature,* vol. 325, no. 6099, pp. 31–36. For a brief but comprehensive review of fossil history, turn again to Campbell's *Humankind Emerging* and also the texts listed later in this note. We learn, for example, that "Neander" was the Greek translation of the name of a gentle, hymn-writing scholar who once hung out near the cave and whose name, interestingly enough, meant "New Man." The history of Neandertals and the other cave types comes, among other sources, from "The Neandertals and Their Evolutionary Significance: A Brief Historical Survey," by Frank Spencer, in a splendid if highly technical compendium called *The Origins of Modern Humans: A World Survey of the Fossil Evidence,* edited by Spencer and Fred H. Smith (New York: Alan R. Liss, 1984). This book is a key reference work and will be used throughout.

The Schaaffhausen quote can be found in an article entitled "On the Crania of the Most Ancient Races of Man," by Schaaffhausen with remarks by Huxley in Huxley's 1863 book *Evidence as to Man's Place in Nature*. The piece is reprinted in an absolutely invaluable collection of essays edited by L. S. B. Leakey and entitled *Adam or Ape* (Cambridge: Schenkman Publishing, 1971). See, on p. 159, Schaaffhausen's own account of the discovery, along with excerpts from a Fuhlrott letter and the Huxley commentary.

The nasty ways of paleoanthropologists, their endless controversies and splitting of hairs, are also alluded to in one of the sources for the quote from Boule, a 1979 article in *Scientific American* by Erik Trinkaus and W. W. Howells entitled, "The Neandertals," 241, (6):118–133. The actual quote from Boule comes from his short 1908 note to the French Academy of Sciences reprinted in the Leakey book.

One finds other little nuggets of interesting (and often absurd) facts in helpful general references that I will quickly spill here to clear the way: C. Loring Brace and Ashley Montagu's *Human Evolution*, 2nd edn. (New York: Macmillan, 1977), which tells how Neandertal was also hypothesized to have been "a rickety Mongolian Cossack, a residue of the Russian forces that had chased Napoleon back to France in 1814" (p. 190); Clifford J. Jolly and Fred Plog's *Physical Anthropology and Archeology* (New York: Alfred A. Knopf, 1979); Clifford D. Simak's *Prehistoric Man* (see especially "Apology to Neanderthal") (New York: St. Martin's Press, 1971); and last but hardly least, the November 1985 issue of *National Geographic*, which has the best introduction to hominid lineages that a novice can buy (along with very clever illustrations).

The Huxley quote comes from his *Evidence as to Man's Place in Nature*. The "knock-kneed" quote is cited, as Spencer points out, in T. D. McCown and K. A. R. Kennedy, *Climbing Man's Family Tree: A Collection of Major Writing on Human Phylogeny, 1699–1971* (Englewood Cliffs, N.J.: Prentice-Hall, 1972), for anyone with a good library within ready reach. According to Brace, by the way, Haeckel had a nearly mystical faith in the superiority of the German "Volk," was a passionate supporter of Bismarck, and helped establish the Monist League, a pseudoreligion incorporating evolutionary ideas that transformed into the rising phenomenon of the Nazi movement. Although we won't name names, a later foundation sponsoring anthropological work in Europe may also have had its share of Nazis. The 1896 quote from Dubois is from Dubois's paper reprinted in Leakey, ed., *Adam or Ape*, p. 168, as is the "diseased man" quote, this time on p. 169. I should point out here the subsequent speculation that the Java femur was actually from a modern human and not from *erectus* after all, though, from other more complete fossils since, we know *erectus* was basically what we thought it was, certainly of the genus *Homo,* standing on two feet and modern in many bodily ways.

The account of Taung is from innumerable sources. For the latest dispute see a letter published by the omnipresent Milford Wolpoff, along with Janet Monge and Michelle Lampl, in *Nature,* vol. 335, October 6, 1988, p. 501. I could (but don't) spell the place of discovery "Taungs" (as opposed to the more popular singular "Taung") in deference to the plural spelling of the place used by the late Dart himself in his *Nature* paper (vol. 115) of 1925.

CHAPTER 3

For a good explanation of mitochondrial DNA, see J. N. Spuhler's "Evolution of Mitochondrial DNA in Monkeys, Apes, and Humans," in the *Yearbook of Physical Anthropology,* 31:15–48 (1988). Also, see Doug Wallace's "Mitochondrial Genes and Disease," in *Hospital Practice,* October 15, 1986. Obviously, most of the chapter comes from the *Nature* paper, vol. 325, no. 6099, pp. 31–36. Other information comes from interviews with the geneticists and telephonic chats with *Nature* editors in London and Washington. Still other material is from a speech Wilson gave in 1989 at a meeting of the American Association for the Advancement of Science.

The gene metaphor is borrowed from Edward Frankel's *DNA: The Ladder of Life,* 2nd edn. (New York: McGraw-Hill, 1979). For references on some of the history of paleoanthropology, see the general references cited in the note above. Howells, who was one of the Berkeley consultants listed in the paper, is author of *Mankind in the Making,* rev. edn. (Garden City, N.Y.: Doubleday, 1967), see pp. 244–247. So was Ernst Mayr a consultant listed in *Nature.* I borrow the term "Garden of Eden" school of thought from paleoanthropologists, using it instead of the older, stodgier, and less clever nickname of "Noah's Ark." I am not one to mix biblical metaphors, and they are the same thing in most regards. For an example of Weidenreich's detailed approach, confer with "Giant Early Man from Java and South China," vol. 40, part I, *Anthropological Papers of the American Museum of Natural History* (New York, 1945). The Cleopatra quote comes from Cann's article "DNA and Human Origins" in the 1988 *Annual Review of Anthropology,* 17:130.

"Paucity of fossils . . . " is from Howells's paper, "Explaining Modern Man: Evolutionists versus Migrationists," *Journal of Human Evolution,* vol. 5 (1976), p. 493. "Dioramas . . . " is from Cann's piece "The

Mitochondrial Eve" in the Natural Science section of the magazine *The World and I* (September 1987), p. 257. "Besides the likelihood . . . " is from Cann's excellent article "In Search of Eve" in the September–October 1987 issue of *The Sciences,* p. 30. "As molecular biologists . . . " is in *The World and I,* p. 259. Wilson's "Some people don't like . . . " comes from *Science,* vol. 237, September 11, 1987, p. 1292. His "set of data" quote, meanwhile, is from John Gribbin and Jeremy Cherfas's book on primate evolution, *The Monkey Puzzle* (New York: Pantheon, 1982), p. 236.

CHAPTER 4

Washburn's quote on the immunological concept is from his article "The Evolution of Man," *Scientific American,* 239 (3). The 1962 paper is Pauling and Zuckerkandl's "Molecular Disease, Evolution, and Genetic Heterogeneity," in Michael Kasha and Bernard Pullman, eds., *Horizons in Biochemistry* (New York: Academic Press, 1962); see especially pp. 200 and 201. Most mutations, they point out, are detrimental to health. On p. 190 is an eye-popper: "The appearance of the concept of good and evil, interpreted by man as his painful expulsion from Paradise, was probably a molecular disease that turned out to be evolution." What! The two men had been looking at molecular changes that cause ailments such as sickle-cell anemia.

The 1965 paper is Zuckerkandl and Pauling's "Evolutionary Divergence and Convergence in Proteins," in Vernon Bryson and Henry J. Vogel, eds., *Evolving Genes and Proteins* (New York: Academic Press, 1965); see especially p. 148. The Wilson and Sarich quote on a 5-million-year-old ancestor is contained in the paper "Immunological Time Scale for Hominid Evolution," *Science,* vol. 158, 1967, pp. 1200–1203. See also *Science,* vol. 154, December 23, 1966, pp. 1563–1566, and *Proceedings of the National Association of Sciences,* vol. 58 (1967), starting p. 142. Also, *Science,* vol. 179, pp. 1144–1147 (generation time in primates); and *Biochemical Genetics,* 7:205–212 (1972).

"The capacity of molecular taxonomy . . . " is from the Washburn article, "The Evolution of Man." Wilson's claim that anthropology is not quantitative or objective is from Gribbon and Cherfas's *The Monkey Puzzle,* p. 252. Meanwhile his complaint about being ignored is in Roger Lewin's dependable book on fossil controversies, *Bones of Contention,* (New York: Touchstone, 1988), p. 116. The Buettner-Janusch quote is from a paper with Robert L. Hill in "Molecules and Monkeys," *Science,* vol. 147, February 1965, p. 836. The bizarre account of his poison candy can be found in *The New York Times*, July 15, 1987, p. B1. His "No fuss, no muss" quote is from Lewin's *Bones of Contention*, p. 112. The Sarich quote about fossils older than 8 million years is from *Background for Man*, edited by Sarich with P. J. Dolhinow (Boston: Little, Brown, 1971). The Wilson complaint about Goodman is from the Lewin book, p. 110. Meanwhile, Sarich's attack on Goodman is from Sarich's "Retrospective on Hominid Macromolecular Systematics," in a review of the molecular clock edited by J. E. Cronin, p. 143. Wilson penned a lengthy review, "Biochemical Evolution," which gives a candidly opinionated perspective on the protein work, in the *Annual Review of Biochemistry*, 46:573–639 (1977). See also for the mammoth albumin, *Science*, vol. 209, July 11, 1980, pp. 287–289. In *Natural History* (December 1973), pp. 72–73, Sarich discusses the panda. For a layman's entrance to DNA, see Maitland A. Edey and Donald C. Johanson, *Blueprints* (Boston: Little, Brown, 1989).

CHAPTER 5

My explanation of the mitochondrion comes in part from *Mitochondria*, by Alexander Tzagoloff (New York: Plenum Press, 1982). See also the *Yearbook of Physical Anthropology*, 31:15–48 (1988), and Douglas C. Wallace in *Trends in Genetics*, vol. 5, no. 1 (January 1989). Mitochondrial DNA codes for about 10 percent of the mitochondrion's own proteins plus twenty-two tRNAs, and a large and small rRNA. The nuclear DNA, demonstrating its importance, codes for the rest of the mitochondrion's proteins. The quote about the "more straightforward manner" of mtDNA is from Brown's chapter in Masatoshi Nei and Richard K. Koehn, eds., *Evolution of Genes and Proteins* (Sunderland, Mass.: Sinauer Associates, 1983), p. 63.

The 1974 paper is "Restriction Endonuclease Cleavage Maps of Animal Mitochondrial DNAs," *Proceedings of the National Academy of Sciences USA*, vol. 71, no. 11 (November 1974), pp. 4617–4621. For Cavalli-Sforza's initial findings, see "Analysis of Human Evolution," in *Genetics Today, Proceedings of the 11th International Congress of Genetics*, The Hague, The Netherlands, ed. S. J. Geerts (New York: Pergamon, 1963), no. 3, pp. 923–933, a paper by Cavalli-Sforza and A. W. F. Edwards. Also, *American Journal of Human Genetics*, 19:233–257 (1967). For more on Brown's work, see D. J. Cummings, P. Borst, I. B. Dawid, S. M. Weissman, and C. F. Fox, eds., *Extrachromosomal DNA* (New York: Academic Press, 1979), pp. 485–499.

The Leakey announcement was in *Nature*, vol. 242, April 13,

1973; pp. 447–450. Brown's 1980 paper was "Polymorphism in Mitochondrial DNA of Humans as Revealed by Restriction Endonuclease
Analysis," *Proceedings of the National Academy of Sciences USA,* vol. 77,
no. 6 (June 1980), pp. 3605–3609. See also "Mitochondrial DNA Sequences of Primates: Tempo and Mode of Evolution," by Brown, Ellen
M. Prager, Alice Wang, and Wilson, *Journal of Molecular Evolution,* 18:-
225–239 (1982); and Brown, Matthew George, and Wilson, "Rapid Evolution of Animal Mitochondrial DNA," *Proceedings of the National Academy of Sciences USA,* vol. 76, no. 4 (April 1979), pp. 1967–1971. One
textbook I used is Charlotte J. Avers, *Genetics* (New York: Van Nos-
trand, 1980).

CHAPTER 6

Cann's 1984 paper is *Genetics,* 106:479–499 (March 1984). Obviously
lab protocol varies a bit from laboratory to laboratory, and changes over
time. I aimed for the essence of it. I should add that the characterizations
in this chapter do not all come from Cann by any means. For technical
information on the entire mitochondrial genome, consult "Sequence
and organization of the human mitochondrial genome," *Nature,* vol.
290, April 9, 1981, pp. 457–464. For a good overview of DNA, consult
the *Scientific American* book entitled *Genetics,* with an introduction by
Cedric I. Davern. Articles I used included ones by Vernon M. Ingram
and F. H. C. Crick. There is good background too in Robert A. Wein-
berg's "The Molecules of Life," *Scientific American,* vol. 253, no. 4 (October 1985), pp. 48–57. Cherfas wrote *The Monkey Puzzle* with John Gribbin, quoting Wilson on pp. 253–254.

CHAPTER 7

"Ours is the most comprehensive . . . " is from the same *Genetics* paper
quoted in the previous chapter note, p. 488. Wilson's contention about
"the molecules of life" is in his 1985 article in *Scientific American,* "The
Molecular Basis of Evolution," 253(4): 164–173. The paper in the Liss
publication is Cann, Brown, and Wilson's "Evolution of Human Mitochondrial DNA: A Preliminary Report" in B. Bonne-Tamir, T. Cohen,
and R. M. Goodman, eds., *Human Genetics, Part A: The Unfolding Genome*
(New York: Alan R. Liss, 1982), pp. 157–165. The paper on length
mutations is Cann and Wilson, "Length Mutations in Human Mitochondrial DNA," *Genetics,* 104:699–711. The 1985 paper is the *Biological
Journal of the Linnean Society,* 26: 375–400; other authors included G.

Higuchi and Ellen M. Prager. Let me point out here that the 2 percent/ million years rate is for both lines, not one. It would be half that, or a 1 percent rate, for a single evolving line.

The *Reader's Digest* story and description of Ames was from a condensation of an article in the August 1987 issue of *California Magazine.* The 1986 paper is Stoneking, K. Bhatia, and Wilson's "Mitochondrial DNA Variation in Eastern Highlanders of Papua New Guinea," in D. F. Roberts and C. F. DeStefano, eds., *Genetic Variation and Its Maintenance* (Cambridge: Cambridge University Press, 1986), pp. 87–100. The Cold Spring Harbor paper is "Rate of Sequence Divergence Estimated from Restriction Maps of Mitochondrial DNAs from Papua New Guinea," *Cold Spring Harbor Symposia on Quantitative Biology,* 51:433–439, Cold Spring Harbor Laboratory, 1986. The money figures come from calls and Freedom of Information requests to the NSF and NIH. Cann's "We assume . . . " is on p. 260 of *The World and I* paper previously cited. For the possibility of 39 percent white input in black genes, see James Spuhler in the *Yearbook of Physical Anthropology,* 31:- 15–48 (1988). But also note that Cann's African sample showed only twenty-eight different morphs that are common to her African and Caucasian samples, but thirty-seven different morphs that are common to her African and Asian samples: evidence, says Spuhler, against a considerable white female mixture in Cann's specific sampling.

The nuclear DNA analysis is "Evolutionary relationships of human populations from an analysis of nuclear DNA polymorphisms," by Wainscoat, Hill, Boyce, Flint, Hernandez, Thein, Old, Lynch, Falusi, Weatherall, and Clegg, in *Nature,* vol. 319, February 6, 1986. Pertaining to New Guinea and Australia settlements: the other option, said Stoneking, was that subsequent migrations through New Guinea that didn't proceed as far as Australia had obliterated pockets of population which indeed would have shown a connection between the two large islands. In other words, ancestral connections between the two may have been replaced by later colonizing types, obfuscating the issue.

CHAPTER 8

The *Chronicle* reporter was a good one, Charles Petit. Had the *Times* decided the story was newsier than it did, it would have placed it somewhere in the A section or on the front of its Tuesday science section— and that would have guaranteed a true run of major publicity. See the *Journal of Molecular Evolution,* 20:99–105 (1984), for Avise paper on lineage survivorship. It should be noted that Wes Brown had mentioned, in his 1980 paper, the fact that his data could be interpreted other ways, but this qualification had been deleted by an editor. The first *Nature* mention of this "Eve," from what I can tell, was an article by Nick Barton and J. S. Jones (vol. 306, November 24, 1983), who reviewed some of the early mitochondrial DNA work and concluded: "The matrilinear patterns of mitochondrial inheritance do mean that we are all more closely related to our mothers than our fathers: to Eve than to Adam. What will sociobiology make of this?" Wainscoat's pointed review is in *Nature,* vol. 325, January 1, 1987, p. 13.

One of the most recent and comprehensive overviews of *erectus* is G. Philip Rightmire's "Homo Erectus and Later Middle Pleistocene Humans," *Annual Review of Anthropology,* 17:239–59 (1988). The figure of 1.3 million *erectus* was given me by Rightmire, who found the information in a *Human Biology* paper (December 1984) by Ken Weiss. In his previously cited paper, Avise quotes only 125,000 for the Lower Paleolithic. The "if this were the case . . . " quote about development into modern *Homo sapiens sapiens* in more than one place is Wilson, Stoneking, Cann, Prager, Ferris, L. A. Wrischnik, and R. G. Higuchi, "Mitochondrial Clans and the Age of Our Common Mother," in F. Vogel and K. Sperling, eds., *Human Genetics* (Berlin: Springer-Verlag, 1987), p. 163.

CHAPTER 9

The Darwin quote from 1871 is on p. 520 of the famous (and infamous) book; a portable version of both that work and *The Origin of Species* is put out by Random House's Modern Library. It's interesting how many geneticists are starting to second-guess natural selection. The description of *Aegyptopithecus* (which was contemporary with another ape called *Propliopithecus*) comes from a conversation with Simons. It is fascinating to wonder if, because *Aegyptopithecus* is older than any monkey teeth found (the oldest being 15 to 16 million years, found at Lake Victoria), apes were ancestral to monkeys. Simons too wonders but doesn't think so. Yet he asks: "Where were the monkeys in Africa during the early Miocene?" Pickford's quote is taken from "Kenyapithecus: A Review," in Phillip Tobias, ed., *Hominid Evolution* (New York: Alan R. Liss, 1985), pp. 107–108. The discussion of *Proconsul's* physical characteristics is partly from the most recent account of *Proconsul,* which can be found in the January 1989 issue of *Scientific American,* pp. 76–82, and also from Pilbeam's article, "The Descent of Hominoids and Hominids," *Scientific American* (March 1984), pp. 84–96.

Caution on fossil bones: one time crocodile thighbones were

mistaken for *Proconsul* collarbones! I had the opportunity to hold Leakey's *Proconsul* skull in my hands during a visit to the National Museums of Kenya in 1988. The crushed skull is small enough to belong to a terrier dog. Pilbeam, in a 1986 *American Scientist* article (vol. 74), called the new view of ape phylogeny a "belated vindication for Haeckel, whose diagram of the hominoid family tree, published in 1866, is remarkably similar to what most researchers believe today. Thus far have we come." A technical discussion of the continental upheavals can be found in *Ancestors: The Hard Evidence* (New York: Alan R. Liss, 1985), pp. 42–50. The most detailed observations of *Gigantopithecus*'s teeth, by the way, are in Franz Weidenreich's "Giant Early Man from Java and South China," *Anthropological Papers of the American Museum of Natural History*, vol. 40 (1945), pp. 63–94. (Weidenreich believed this huge ape to be a hominid. Some think it may have been twelve feet tall!) For a discussion of "bigfoot," see *Nature*, vol. 313, February 7, 1985, p. 418, vol. 304, July 14, 1983, in the book review section, and vol. 320, March 13, 1986, p. 104. Or *Discover* (March 1988), pp. 45–53.

For a discussion of whether the orangutan is closer to man than now thought, see *Nature*, vol. 308, April 5, 1984, p. 501. Louis Leakey, interestingly enough, thought *Kenyapithecus wickeri* was already making use of stones to smash skulls and bones to get at brain and marrow (see p. 445 of *Adam or Ape*). His son Richard told me he no longer considers *Ramapithecus* to be represented in the African fossil record and is not very certain *Kenyapithecus* is on the direct human line. He also emphasizes that another ape, *Afropithecus*, is earlier still, going back 18 or 19 million years, with morphological characteristics that apparently link it to *Ramapithecus*. Leakey (see *Nature*, vol. 324, November 13, 1986, p. 143) says: "The genus *Sivapithecus* has been reported from East Africa but the new material supports the hypothesis that this genus only occurs beyond Africa."

The molecular relationships between apes and man are summarized on p. 150 of Peter Andrews's paper, "Molecular Evidence for Catarrhine Evolution," in Bernard Wood, Lawrence Martin, and Peter Andrews, eds., *Major Topics in Primate and Human Evolution* (Cambridge: Cambridge University Press, 1986). For Sibley and Ahlquist's 6.3 to 7.7 million figure, see "The Phylogeny of the Hominoid Primates, as Indicated by DNA-DNA Hybridization," *Journal of Molecular Evolution*, 20:2–15 (1984). See also Sibley and Ahlquist's "Reconstructing Bird Phylogeny by Comparing DNA's," which is in *Scientific American* for February 1986 and is written in more comprehensive prose than the work of other DNA evolutionists like Wilson. Goodman's views can be seen in papers such as those in *Systematic Zoology,* 31:376–399; and *Acta Zoologica Fennica,* 169:19–35.

Further discussion: Philip D. Gingerich's "Nonlinear Molecular Clocks and Ape-Human Divergence Times" in Tobias, ed., *Hominid Evolution,* cited above. The "erratic and sloppy" quote is p. 4 of the Sibley and Ahlquist paper on man and apes cited above. Sarich's "so captured our attention . . . " (in discussing Sibley and Ahlquist's finding) is Sarich, Carl Schmid, and Jon Marks, in "DNA Hybridization as a Guide to Phylogenies: A Critical Analysis" (unpublished at time of writing). Eventually a version of this was published in the *Journal of Human Evolution,* where both Sarich and Marks serve as associate editors. Information is also drawn (especially the "magnificent achievement" quote) from Sarich's "An Independent Look at the Sibley-Ahlquist DNA Hybridization Results and Conclusions," which was unpublished when I received it from Sarich at the Asilomar seminar in California. Simons, who, as the champion of *Ramapithecus,* had the most to lose (along with Pilbeam, his former partner) from the episode, now accepts a recent split between chimps and apes and the disqualification of *Ramapithecus* as a hominid but still holds out the possibility that it may one day be reinstated in man's lineage, bristling at the mention of the Berkeley group. "I mean, it may make Sherry Washburn happy if he can now bask in retirement and say, 'I was the one who first urged that the split between man and the African ape was very recent,' but unless he did it on some kind of factual evidence, what was it?" says Simons. "It was nothing but a prophecy that turned out to be right. I believe Nostradamus made so many that he's been discovered to be right a lot of the time."

Goodman's findings on nucleotide sequencing can be seen in a paper by Goodman, Michael M. Miyamoto, Ben Koop, Jerry Slightom, and Michele Tennant, "Molecular Systematics of Higher Primates: Genealogical Relations and Classification," *Proceedings of the National Academy of Sciences USA,* vol. 85 (October 1988), pp. 7627–7631. Zuckerkandl's view was taken from *Science,* vol. 241, September 23, 1988, p. 1598, while the opinion on whether Wilson acted properly is from the same journal, September 30, 1988, p. 1757. Marks even attacked the writer of the *Science* articles, biochemist Roger Lewin, claiming Lewin gave Sarich and Marks a negative slant because Marks had written a slightly critical review of a Lewin book. It never stops. Ironically, no science writer had been more receptive to Sarich and Wilson's work than Lewin. I use "hominid," by the way, for creatures that were not apes but also were not quite human. A "hominoid," on the other hand, is the great family that supposedly includes both apes and humans. As for the similarities between man and chimp, after spending part of the afternoon playing with one named Charlie (who knew karate), I came away with fond feelings for the little fellow, who lives in Niagara Falls, but with the definite conviction that no matter what a DNA sequence says, there's one heckuva lot more than a 1 percent difference between Charlie and humans. During one debate between Sarich and a creationist scientist named Duane Gish, Gish retorted that both man and watermelons are composed of about 97 percent of water but that doesn't make them very close.

CHAPTER 10

See references in later chapters for Nei. For oldest hominids, see F. C. Howell's monograph in Vincent Maglio and H. B. S. Cooke, eds., *Evolution of African Mammals* (Cambridge, Mass.: Harvard University Press, 1978), pp. 155–233. For an alternate view of knucklewalking, see Alan R. Templeton's "Phylogenetic Inference from Restriction Endonuclease Cleavage Site Maps with Particular Reference to the Evolution of Apes and Humans," in *Evolution*, vol. 37, no. 2 (March 1983). You say there were no man-apes in Asia? See the collection of papers *Hominid Evolution*, previously cited, pp. 255–263, the chapter by Jens Lorenz Senckenberg. Boy, do the teeth of *Hemanthropus peii* from China look like a robust man-ape's! (But then there have been a few instances where teeth resembling australopithecines have been noticed in naval recruits, and as Milford Wolpoff says, "I know of no australopithecines in the Navy.") Here's more food for thought: is it true *boisei* had some teeth that looked a lot like *Gigantopithecus*, the "abominable snowman" from Asia? How could anybody be sure of very much when, as Johanson mentioned, at the time of Lucy there still had not been discovered a single fossil chimpanzee skull? There are those who say tool use went back as far as 3 million years ago, and that australopithecines were hunters (one early adherent is Sherwood Washburn; and though new discoveries such as tooth microwear may have changed many of the views expressed in Robert Ardrey's *The Hunting Hypothesis* [New York: Atheneum, 1976], some of the rationale arguing against scavenging should perhaps be reconsidered since it is powerful enough to still have the ring of some percentage of truth). Another cautionary note: paleoanthropologists and paleontologists battle it out over environmental factors. Pilbeam has challenged a popular view that worldwide climatic change 2.4 million years ago spurred evolution of robust australopithecines. How can this be, asks Pilbeam, when *boisei* in the form of the "black skull" goes back to 2.5 million years? (For an alternate view, see E. S. Vrba, pp. 63–71, in *Ancestors: The Hard Evidence.*) See also Karl W. Butzer in *American Scientist* (September–October 1977), pp. 572–584. There are even those (Marc Verhaegen, *Nature*, vol. 325, January 2, 1987, p. 305) who don't believe our hominid "ancestors" lived in the savannahs, for we have a water-and-sodium-wasting cooling system, it is argued, that is completely unfit for a dry environment.

On matters of age, there are also those who believe the Taung child might have been only 1.1 million years old. Or younger. By the way, Broom had his many magic moments too, once happening upon a skull that glistened like a diamond because of the encrusting lime crystals reflecting the right rays of sun. Raymond Dart, meanwhile, died on November 22, 1988, nearly sixty-four years to the day since receiving a chunk of sandy limestone containing the Taung skull. For study of heel bones, see "Comparative Study of Calcanei," in Tobias, ed., *Hominid Evolution*, p. 247. As for Mary Leakey: I use her own book, *Disclosing the Past* (Garden City, N.Y.: Doubleday, 1984), for some of the background, as well as an interview during which she expanded upon her first trek to Laetoli:

> We were working at Olduvai and a Masai came to us and told us that he knew another place where you got fossil bones like those at Olduvai. In fact he went himself by foot and fetched some fossils from Laetoli and showed us. We then went across, and it was a long safari in those days. There were absolutely no roads, and it took us, I think, three days to get there going via Lake Ndutu. It was a long way, and we had a very old vehicle, but we got there. A Masai showed us the way across the plains. There was material that was so broken up that we only spent a month or six weeks there and then returned to Olduvai, because we had no idea of the dates.

Actually, Louis had found a hominid tooth at Laetoli—the first adult *Australopithecus* ever, preceding even Broom—but it had been misinterpreted as belonging to a monkey and was only recognized as a hominid much later. Meanwhile Johanson's opinion about Mayr being the "last court of appeal" is in the contentiously and engagingly written *Lucy*, by Johanson and Maitland Edey (New York: Warner Books, 1982), p. 290. *Lucy* captures wonderfully the moment of the dramatic discovery: "I can't believe it," [Johanson] said. "I just can't believe it" (p. 17). Johanson and his co-discoverer, Tom Gray, began hugging each other and howling. "We've got to stop jumping around," [Johanson] finally said. "We may step on something." (Johanson penned a follow-up book, *Lucy's Child,* in late 1989—New York: William Morrow.)

The quotes here from Richard Leakey are from an interview in Chicago, while other quotes further on in the book are from an interview at the National Museums of Kenya, which Richard quit early in 1989 during a dispute that had partly to do with government inaction on elephant poaching. But he was back on the job—at the museum he did so much to build up—by the beginning of February and then was made wildlife commissioner! I am grateful to Royce-Carlton, a lecture agency Leakey and I share, for arranging our initial session. Mary was gracious enough to drive through midday traffic to meet with me. She strongly dislikes speaking with journalists and as a schoolgirl once hid in a boiler room to avoid reading aloud in a poetry class. For years she lived alone at Olduvai, savoring the utter solitude of the environs there.

Mayr's quote on all the "bewildering" types is on p. 231 of *A History of American Physical Anthropology, 1930–1980* (New York: Academic Press, 1982). Cann's ending quote is from the previously cited piece in *The Sciences,* p. 32. Before elaborating on the *afarensis* controversy, let me cite a few more helpful references, especially for the *Australopithecus* buffs: for one discussion of the age of Hadar beds, see *Nature,* vol. 313, January 24, 1985, p. 306. For reconstruction of diets of more recent primitive Africans, see *Nature,* vol. 319, p. 321. For discussion of *Australopithecus* diet based on tooth microwear (and finding that *robustus* ate coarser food than earlier hominids), see Frederick E. Grine and Richard F. Kay in *Nature,* vol. 333, June 23, 1988, pp. 765–768. See also Glynn Isaac's short article, "The emergence of man" in *Nature,* vol. 285, p. 72.

The debate on what various man-apes ate is a can of worms I've tried not to open. "Microwear can't tell you exactly what they were eating because you can eat a sand grain and have a pit on your tooth or eat a nut and have a pit on your tooth or oysters with shell and have a pit on your tooth," Frederick Grine of Stony Brook explained to me. He warns that when it comes to estimating body sizes, "some people are just making it up out of whole cloth." You may find a big femur but not know which skull it belongs to. He has a new volume of papers out on australopithecines, published by Aldine.

Details on recent discoveries in the Middle Awash can be found in *Nature,* vol. 307, February 2, 1984, starting on p. 423. See *Scientific American* for a recent (November 1988) overview and theorizing on bipedality. Robust toolmaking arguments can be glimpsed in Randall L.

Susman's "Hand of *Paranthropus robustus* from Member 1, Swartkrans: Fossil Evidence for Tool Behavior," *Science,* May 6, 1988, pp. 781–783. See Johanson's response to technical dispute over how many types of hominids there were at the Afar in "Craniodental Morphology of the Hominids from Hadar and Laetoli: Evidence of *'Paranthropus'* and *'Homo'* in the Mid-Pliocene of Eastern Africa?" in the previously cited *Ancestors: The Hard Evidence.* From that book, see too Todd R. Olson's paper disputing what Johanson just said (pp. 103–117), and for environmental details of Kenya, see *Earliest Man and Environments in the Lake Rudolf Basin* (Chicago: University of Chicago Press, 1976), which includes as editors Richard E. F. Leakey and Glynn Isaac, two who certainly knew the terrain. As for the criticism from Harvard: "Ernst Mayr is a little bit of an enigma," Johanson responds. "He initially wrote to us and was fully behind the new name [*afarensis*]. I can show you the letter. He said he saw nothing wrong with it. Why he's changed his mind I have no idea. We could have called it anything we wanted to. There's nothing in the code [of nomenclature] that prevents you from doing that. Mary wanted some of the fossils called *Homo* because she didn't like the name *Australopithecus.* The problem is that from an historical perspective we're stuck with it. I don't like it either. Because what it means is 'southern ape.' "

In simpler terms the Leakeys, though they recognized Lucy as a man-ape, did not believe all the other fossils found in the Afar, representing at least 13 other hominids, were australopithecine but rather that some of the fossils, a windfall of more than 200 bones and teeth, represented ancient *Homo.* The range of time for these fossils was roughly 2.8 million to 3.6 million years old, overlapping or more likely slightly younger than what was found at Laetoli. Forget for a moment about the name. Johanson and White were also, in the view of their rivals, fooling around with the evolutionary tree, putting their Lucy-type man-apes as the oldest direct human ancestors and shunting aside the Taung child and other man-apes of that australopithecine species known as *africanus* onto a dead evolutionary branch. I have no idea where *africanus* will end up, but there is the intuition (didn't it look like some habilines? and in linking *afarensis* with *habilis* to the exclusion of *africanus,* didn't that imply, as Henry McHenry and Randall Skelton point out, parallel or convergent evolution of forty-two traits?) that it was not so dead a branch, and I don't know how many species were at Hadar but I do know that the time has about come for two talents like Leakey and Johanson to bury the hand ax (see next chapter) and once more call each other friends.

CHAPTER 11

Louis Leakey's announcement of *habilis* in *Nature* in April of 1964 was co-authored by Phillip Tobias and J. R. Napier. Responding to those who doubted and still doubt whether the new discovery deserved a new species label, Mary says, "There is such a thing as *habilis*. Of that I have no doubt at all." Could robust man-apes also have been toolmakers? "Oh, they could, yes," says Mary. "I just don't know." There are those who doubt the circle of stones was a rudimentary shelter, instead believing them to be rocks pulled up in a strange pattern by a tree. And others dismiss the notion that *habilis* was a hunter. What Cann has to say about *habilis* is from her piece in *The Sciences*, p. 32. For meat-eating see letter page of *Nature*, vol. 325, January 22, 1987. See also "Archeological evidence for meat-eating by Plio-Pleistocene hominids from Koobi Fora and Olduvai Gorge," by Harry T. Bunn, in *Nature*, vol. 291, June 18, 1981, pp. 574–576.

On brain advancement, see John E. Yellen's "News and Views" piece in *Nature*, vol. 322, August 28, 1986, p. 774. Holloway told me that Tobias is wrong to think the Wernicke area of the brain is also enlarged in *habilis*. The figure 110 hominids within five years is from a chapter Leakey wrote in *Earliest Man and Environments in the Lake Rudolf Basin*, p. 476. The issue of the *American Journal of Physical Anthropology* with the new finds from Leakey was 67:135–163 (1985). In his book *People of the Lake* (New York: Avon, 1978), it is stated that by the end of the first decade at Turkana, "fragments of fossilized bone from several hundred prehuman creatures" had been unearthed (p. 9). Leakey's "This was fantastic new information" is from p. 149 of his autobiography *One Life* (New York: Salem House, 1983), a book that is hard to put down. The Wood synopsis can be found in his chapter "Early *Homo* in Kenya, and Its Systematic Relationships" in the previously cited *Ancestors*. Leakey's quote on *habilis* living at the same time as early australopithecines is on p. 77 of his book *Origins* (New York: E. P. Dutton, 1977). The remembrance of his father's reaction is from p. 150 of *One Life*. The initial 2.6 million date was eventually adjusted to 2.42 before the final analysis.

There is no better rendition of the KBS controversy than Roger Lewin's *Bones of Contention*, previously cited, which is where I first learned of the *Time* sales. A spokesman for *Time* told me the Leakey cover ranked up with a historic theological story, "Is God Dead?" (That was in irreverent times spawned by the bad old sixties.) I also draw from Richard Leakey's *The Making of Mankind*, another Dutton book, 1981. The comment about a "lightweight book" is from Mary's autobiography,

p. 184. It is fascinating to note how geochronologists—using even different techniques, like fission tracking—come up with dates that fit in with their previous argon results, while others find the opposite and in the end demonstrate that the dates from both techniques were wrong. Like anyone else, technicians often get the results they want to get, perhaps because of something working in the subconscious during the interpretation and analysis stage.

For Mary's disbelief at how Johanson had reacted to his discovery, see p. 183 of her autobiography. For Johanson and White's new *habilis*-like humanoid, see "New partial skeleton of *Homo habilis* from Olduvai Gorge, Tanzania," *Nature*, vol. 327, May 21, 1987, p. 205. On classifications, Richard Leakey points out that some of the fossils labeled *africanus* in South Africa bear striking resemblances to some fossils currently classified *habilis*. He believes fossil 1813 is like specimens at Olduvai, meanwhile, that are labeled OH13, OH24, and OH62 (the latter being the "*habilis*" fossils found by White in 1986). This category is still in search of a species name. "The real problem," Mary says, "is whether these very small-brained creatures that were contemporaneous with *habilis* were females of *habilis* or whether they are another species. Richard strongly believes they are another creature, and I think he might well be right. OH13 and 1813—I can't really see them as the same as the *Homo habilis* skull that he found at Turkana." If it is given a new taxon, some of the *africanus* and *habilis* fossils may find themselves redesignated to this new category, Richard believes. Leakey has expressed the view that, at the least, gracile australopithecines might better be called human and robust ones labeled as *Australopithecus*. "I would like to see all the hominids put into a single genus *Homo*," he says. "I would think that the line in which the large brain is being developed from around two million years ago may be traceable back well beyond 2.5 million years. Whether it will go back to five million I'm less certain now."

CHAPTER 12

Leakey announced the *erectus* discovery in the August 29, 1985, *Nature*, vol. 316, starting on p. 388. I also draw from Pat Shipman's report starting on p. 94 of the April 1986 *Discover*. For flame use 1.5 million years ago, see *Nature*, vol. 336, December 1, 1988, p. 464. My description of the brutal hunt for gelada baboons comes in part from vol. 22, no. 3 (June 1981), pp. 257–268, of *Current Anthropology*. A date of 600,000 years (actually .6 million to .9 million) was conveyed to me by Pat Shipman of Johns Hopkins. Leakey's quote on self-awareness is from p.

37 of *One Life*. The Wolpoff quote on violence is p. 8 of "Homo erectus and Origin of Human Diversity," in J-J. Hublin and A-M. Tiller, eds., *Aux Origines de la Diversité Humaine* (Paris: Presses Universitaires de France).

The facial changes that came with *erectus* can be studied in greater detail by consulting "Nasal Morphology and the Emergence of Homo erectus," in the *American Journal of Physical Anthropology*, 75:517–527 (1988). The authors point out that in the Nariokotome Boy (KNM-WT 15000), the nasal bones exhibit marked nasal aperture eversion. The reason for the nose developing the way it did may have been to retain moisture from exhaled air as life got dry on the savannah. As for Dubois, in his *"Pithecanthropus erectus*—A Form from the Ancestral Stock of Mankind" (from the Smithsonian translation, 1896), he was more middle-of-the-road. "It is well known that a not inconsiderable number of anatomists and zoologists hold diametrically opposite views regarding the significance of these remains," he wrote shortly after finding the famous skullcap, femur, and teeth near Trinil in central Java. "For instance, as to the skull, a few have believed that it is human, although of much more apelike appearance than hitherto known, while others have considered it the skull of an ape far more human in character than any previously discovered. It is remarkable that only a few have believed in a third possibility, intermediate between these two views, that we have before us here a transition form between apes and men that is neither man nor ape."

Leakey's excited description of finding the *erectus* in 1975 is on p. 172 of *One Life*. I get the idea of faulty geomagnetic interpretation in the case of Yuanmou from Pope in G. Bartstra and W. A. Casparie, eds., *Modern Quaternary Research in Southeast Asia* (Boston: A. A. Balkema, papers read at Symposium 1, 12th Congress of the Indo-Pacific Prehistory Association, Philippines, vol. 9, 1985). For further information on the dates, see Wu Rukang of the Institute of Vertebrate Paleontology and Paleoanthropology, Academia Sinica, Beijing, in *Ancestors*, p. 245. The date of 1.7 million for Yuanmou also comes from that Institute. Pope in "Recent Advances in Far Eastern Paleoanthropology," *Annual Review of Anthropology*, 17:43–77 (1988), says Yuanmou specimens are probably only 500,000 years old. As for the oldest remains in Java, see again Pope's paper in *Modern Quaternary Research in Southeast Asia*. For resemblances of Sangiran mandible, see *Ancestors*, pp. 221–226. Pope's "earliest hominids" quote comes from p. 73 of his latter paper. Cann's quote on seasonal camps is p. 32 of *The Sciences*, previously cited.

Wolpoff's quote on why *erectus* left Africa is from p. 609 of the

November 1985 *National Geographic,* which presented, as I said, among the best synopses and illustrations of hominid stages ever printed. For pigment, see "Hominid occupation of the East-Central Highland of Ethiopia in the Plio-Pleistocene," *Nature,* vol. 282, November 1, 1979, p. 38. Cann expresses her view on the role of glaciers in her *World and I* piece, p. 261. For Campbell quote, see his textbook *Humankind Emerging,* p. 306. I also rely on Campbell for what was found at the Peking Man site. In this field, not even textbooks are sacrosanct. Where in other fields what is printed in such a text usually involves a consensus conclusion—a "fact"—in paleoanthropology there is so little consensus that a textbook can be perceived as the espousal of a particular philosophy instead of a collection of facts. Campbell, however, is about as close as paleoanthropologists come to objectivity.

The Karlskoga comments were reported by Carl-Axel Moberg of the Institute of Nordic Archaeology in a calm, clean, and bucolic city everyone should have the pleasure to visit, Göteborg, Sweden. The "loose collection" issue was mentioned by G. Philip Rightmire in his superb "Homo Erectus and Later Middle Pleistocene Humans," *Annual Review of Anthropology,* 17:239–259 (1988). The pattern of facial breadth I take from a paper by A. Blisborough of the University of Durham and B. A. Wood of the University of Liverpool. On language physiology: what was known about the pharynx was that the roof was long, indicated by the great basion-prosthion distance, or the distance between the pharyngeal tubercle and staphylion. Frontal, temporal, and parietal regions of the brain were less developed. The fig leaf? Try "Archeological Evidence from the Koobi Fora Formation," by G. Isaac, J. W. K. Harris, and D. Crader in the Isaac–Leakey tome *Earliest Man and Environments in the Lake Rudolf Basin* listed above. Definitions of *erectus,* I might add as an afterword, were put forth in volumes that made anything but barber-chair reading. Here's one characterization from Wolpoff: "anteriorly projecting and vertically tall supraorbital tori separated from the frontal squama by a supraorbital sulcus; flattening of the frontal squama broken by a metopic eminence and presence of a sagittal keel; reduced facial prognathism; broadening of the face, and some reduction facial height, holding sex constant; expansion of the midface, zygomaxillary region, and orbital size, with reduction of the lower face and mandible. . . ." (the French volume cited previously, *Aux Origines . . . ,* p. 9). In short, an *erectus* was never going to make it as a Breck model. Though, compared to *habilis,* it was that great leap toward what we know as true humanity, it must be remembered that we are still dealing with a primitive entity. A million years, or a quarter million years, is not yesterday.

There was also continuing debate over what did or did not constitute a taxon. *Heidelbergensis*, as we will see next, was probably more advanced than *erectus*. Conversely, was Bilzingsleben an *erectus?* Just as there had been arguments over the legitimacy of the taxa *afarensis, africanus*, and *habilis*, and confusion over the phylogenetic position of robusts, so too were there voices that questioned whether *erectus* was a definable species or, at the other extreme, debate not over the legitimacy of the taxon but about whether *erectus* should be split into two species— one for the early African specimens, another for the Asian version. In Asia there are legends of isolated bands of *erectus* still hiding in the mountainous woods. No one gives that much credence, however—at least no more than the "abominable snowman."

CHAPTER 13

For the Java fossils, see Yale University Publications in Anthropology, no. 78, 1980. Howells's quote is from "Homo erectus—Who, When, and Where: A Survey," in the *Yearbook of Physical Anthropology*, vol. 23 (1980), p. 7. Meanwhile Rightmire (*American Journal of Physical Anthropology*, 53:225–241) adds jaw samples from Ileret and Koobi Fora near Turkana as possible archaic specimens. For Petralona dates, see too the *Journal of Archeological Science*, 6:235–253 (1979), "The Significance of the Fossil Hominid Skull from Petralona, Greece," which gives a date of 300,000. The 220,000 date on Terra Amata I got from archeologist Lewis Binford; the 400,000 figure came from Campbell. The Wilson quote is from p. 24 of *Science*, vol. 238, October 2, 1987. See, for description, what Rightmire wrote on p. 248 of the *Annual Review of Anthropology*, vol. 17 (1988).

The additional characteristics distinguishing *erectus* from archaics can be seen on p. 203 of Clark Howell's monograph cited previously. For Lainyamok, see *Nature*, vol. 306, November 24, 1983, p. 365, and personal communication. For analysis of bone collagen, see *Nature*, vol. 319, January 23, 1986, p. 321, "Reconstruction of African human diet using bone collagen carbon and nitrogen isotope ratios," by Stanley H. Ambrose and Michael J. DeNiro.

About Spain: Howell doesn't mention that it was one of his colleagues, Robert Ardrey, who helped immortalize the idea of flame-driven animals. On p. 144 of Ardrey's *The Hunting Hypothesis*, he says the evidence Howell found for a fire-drive was "inarguable." For the Nice site (Terra Amata), I rely upon Campbell. Klein's quote on "The rate of artificial change" is from p. 29 of his "The Stone Age Prehistory of

Southern Africa," in the *Annual Review of Anthropology,* 12:25–48 (1983). For Acheulian sites in Russia, see Yakimov's "New Materials of Skeletal Remains of Ancient Peoples in the Territory of the Soviet Union," in L.-K. Könnigson, ed., *Current Argument on Early Man* (London: Oxford University Press, 1980), p. 154. For the last quote of the chapter I rely on my interview with Rightmire in combination with p. 248 of his paper cited above. I also rely in general upon Chris Stringer's "Middle Pleistocene Hominid Variability and the Origin of Late Pleistocene Humans" in *Ancestors,* pp. 289–295. For the Middle Pleistocene I use the period 730,000 to 128,000 years ago, as Stringer suggests.

For European archaics, it would be hard to find better than Stringer's comprehensive paper (written with J. Cook, A. P. Currant, H. P. Schwarcz, and A. G. Wintle) entitled "A Review of the Chronology of the European Middle Pleistocene Hominid Record" in the *Yearbook of Physical Anthropology,* vol. 25 (1982). See also Mann, "Homo Erectus," in Phyllis Dolhinow and Vincent Sarich, eds., *Background for Man.* In addition to encyclopedia entries and general reference, I use John Imbrie and Katherine Imbrie, *Ice Ages* (Short Hills, N.J.: Enslow Publishers, 1979), and consultation with Parker Calkins in the geology department at the State University of New York at Buffalo for ice ages. There have basically been eleven or twelve periods of ice expansion in the past 1 million years, especially in the past 400,000, and geologists are getting away from using terms like "Riss" and "Mindel" glaciations. Things aren't so clear-cut or easily defined. The dates for glaciations have in the past differed by as much as 200,000 years, with geologists using various definitions for a glaciation's starts and stops.

CHAPTER 14

The quote on where the transition occurred is Rightmire's "Patterns in the Evolution of *Homo erectus,*" in *Paleobiology,* 7(2), p. 246. For Cann on Neandertals, see again *The World and I,* p. 263. For the role of southern Russia, see Yakimov's "New Materials of Skeletal Remains of Ancient Peoples in the Territory of the Soviet Union," in L-K. Könnigson, ed., *Current Argument on Early Man* (London: Oxford University Press, 1980), p. 165. For Cann's quote on cranial shapes, see *The Sciences,* p. 33. For Eckhardt's moan, see *Nature,* vol. 326, April 23, 1987. The Japanese researchers expressed themselves on p. 111 of *Nature,* vol. 327, May 28, 1987. For the two Paris researchers, see the same "Scientific Correspondence" page, but in *Nature,* vol. 329. Wilson's smug quote is from *Science,* vol. 237, previously cited, p. 1292.

CHAPTER 15

For Cann's optimism on a good working relationship, see *The World and I*, repeatedly mentioned. Her quote on the "100 percent certainty" that genes have a history is from a paper in the *Annual Review of Anthropology*, 17:127–143 (1988). Weidenreich's comment on Adam and Eve is in a book edited by W. W. Howells, *Early Man in the Far East*, Wistar Institute Press. Coon's thesis quote is from p. 659 of *The Origin of Races* (New York: Alfred A. Knopf, 1973). His outrageous quote on skin color is on p. 317 of *The Living Races of Man* (New York: Alfred A. Knopf, 1965). Another book, *The Story of Man* (New York: Alfred A. Knopf, 1963), preceded these two. Wolpoff's comparison between Australian and Indonesian specimens is from p. 20 of his paper *"Homo erectus and Origins of Human Diversity,"* previously cited and in press for *Aux Origines de la Diversité Humaine*. His, Thorne's, and Zhi's treatise is on pp. 411–483, "Modern *Homo sapiens* Origins: A General Theory of Hominid Evolution Involving the Fossil Evidence from East Asia," in Spencer and Smith, eds., *The Origins of Modern Humans* (Liss).

The Wajak skulls, it could be argued, should be listed separately unless one was going to discuss the mixed Indonesian and Asian origins of Australians. Naturally there were a good number of opponents who felt Wolpoff's linkage of modern Asians to ancient ones was overstated. The traits he cited, asserted Rightmire of Binghamton, "are so general as to imply no special relationships among populations. The fact that facial height and tooth dimensions for one Sangiran individual are similar to Kow Swamp (male) means need not be read to indicate evolutionary continuity over several hundred thousand years. The Ngandong assemblage is claimed to be 'morphologically intermediate' between the Sangiran hominids and modern Australians, but this also can be questioned."

For Cann's view on Asian roots, see p. 263, *The World and I*. For the Wilson quote on paleoanthropologists, see Gribbin and Cherfas, *The Monkey Puzzle*, p. 257. For Wolpoff's quote on "war," see "Multiregional Evolution: The Fossil Alternative to Eden," in *The Human Revolution: Behavioural and Biological Perspectives*, edited by P. Mellars and C. B. Stringer (Edinburgh University Press, in press.) See also "Theories of Modern Human Origins," in Günter Bräuer and F. H. Smith, eds., tentatively *Continuity or Complete Replacement? Controversies in the Evolution of Homo sapiens* (in press); and see Wolpoff's "Human Evolution at the Peripheries: The Pattern at the Eastern Edge," in Tobias, ed., *Hominid Evolution*, the Liss book previously cited, pp. 355–365.

The Aborigines, according to Weidenreich and Wolpoff, shared certain features that were absolutely unique, including a well-developed supraorbital torus which combined a discontinuity at the glabella with the lack of a supratoral sulcus and the long, flat, receding forehead. In some of this, he and Sarich had once shared agreement. Though he still spent much of his day in Wilson's lab, Sarich, for the longest time, maintained many of his previous positions stated within passages of academic text that supported aspects of regional continuity and saw basic truth in Coon's work. Indeed, Sarich had dismissed the idea of a single point of origin for *Homo sapiens*. He posited one theory that had *erectus*-grade human populations evolving in the same evolutionary direction due to the spread of common cultural practices. "In other words," wrote Sarich around the time of Coon's masterwork, "the various human races represent nothing more, in large extent, than different paths along the journey from *H. erectus* to *H. sapiens* grade of evolutionary development. It is not surprising that it is in precisely those parts of the skeleton (skulls, jaws, teeth) that have changed most from *H. erectus* to *H. sapiens* that the greatest differences among human races exist today" (Sarich and Dolhinow, eds., *Background for Man*, p. 107). By February 1989, however, Sarich was dramatically changing his evolutionary theories. For some comments against Wolpoff's morphological evaluations, see Rightmire's paper "Homo Erectus and Later Middle Pleistocene Humans," previously cited, p. 255. See also N. W. G. Macintosh and S. L. Larnach, "Aboriginal Affinities Looked at in World Context," in R. L. Kirk and A. G. Thorne, eds., *The Origin of the Australians* (Australian Institute of Aboriginal Studies, Humanities Press, 1976), in which it is stated that "a comparison of Australian skulls with those of *Homo erectus* clearly shows the complete absence of the *Homo erectus* cranial pattern of the former." For the Sangiran comparisons, I drew from Wolpoff and Alan G. Thorne's "Regional Continuity in Australian Pleistocene Hominid Evolution," *American Journal of Physical Anthropology*, 55:337–349 (1981). The Howells comment was from a phone chat.

CHAPTER 16

For Cann's quote on cultural innovations, see p. 263 of her *The World and I* article. Sarich himself was of two minds. What he had told me in June of 1988 changed radically by early 1989. In a dramatic turnabout, Sarich, believing for the longest time that the first ancestors had left Africa 700,000 to 800,000 years ago, now had a different problem with his colleagues' data: he thought the 200,000 years ago for Eve's existence and 100,000 years ago for the out-of-Africa migration, which he had previously criticized as much too recent, was now too late. Sarich now believed that modern humans probably emerged from Africa as recently as *30,000* years ago, instead of the 700,000 to 800,000 years he had long propounded—a twenty-five-fold change of opinion and a scotching of multi-regional-like theories he had held for years, all in nine months' time. Sarich, after studying a bunch of skulls, had also concluded that Broken Hill was a Neandertal. The Wallace quote is from *Hospital Practice,* October 15, 1986, p. 92. Cann's thinking on Adam comes from p. 37 of her *The Sciences* article. And her jibe at paleontologists and their scientific limits is on p. 257 of *The World and I.*

CHAPTER 17

On Neandertals and Neander: a previous one was actually discovered in 1948 at Forbes' Quarry, Gibraltar, but an analysis was not published until after the far more famous 1856 Neander Valley discovery. Up until a century ago, there were still reports of Neandertal sightings—wild men of the forest!—in parts of southern Europe. One mandible from Bañolas, Spain, has been dated to as late as 17,700 years ago (see *Bulletin of the British Museum of Natural History,* 35, October 29, 1981), but the dating must be in error. If the dating is accurate, however, then obviously Neandertals did not disappear from Europe as suddenly as everyone has thought. And if they persisted into Würm II, said Chris Stringer in 1981, then "these populations could have contributed to the ancestry of a.m. *Homo sapiens* populations of the late Pleistocene." The "pre-*sapiens*" theory is most obviously similar to replacement models, by the way, in that both believed Neandertals and modern man were contemporary. Neandertals lived, some of them at least, until age forty-one (see Trinkaus in *American Journal of Physical Anthropology,* 72:123–129 [1987]). Cann's further remarks on Neandertals are on p. 33 of that brave piece of hers in *The Sciences.*

On similarities in bone structure, see *The Mousterian Legacy,* edited by Erik Trinkaus, BAR International Series 164, 1983. When I say "trivial" differences, I am alluding to overall proportions. The femur shaft breadth index (midshaft breadth/bicondylar femoral length x 100) of a Neandertal can be, say, 7.05 compared to a 5.95 value for moderns (see Table 2 of Smith paper in this volume). Also in this volume is an excellent Clark article on tools, which I draw from in addition to personal communication. Note that the South African blades (Howieson's Port Complex) may go back to 80,000–90,000 and include large lunates and trapezes. Binford, however, says Cann's argument of tools is inaccurate. See also the article by Erik Trinkaus and W. W. Howells, "The Neandertals," *Scientific American,* 241 (6), 1979, which I use as a general reference.

The Holloway statement is on p. 321 of his paper, "The Poor Brain of *Homo sapiens neanderthalensis:* See What You Please . . ." in *Ancestors.* For the effects of increased stature, see pp. 330–331 of *Ancestors,* an obviously invaluable reference. That Neandertals formed a major population, see Trinkaus and Howells's article in *Scientific American,* p. 118. The comment on Vindija Neandertals is on p. 541 of Wolpoff, Smith, Mirko Malez, Jakov Radovcic, and Darko Rukavina in the *American Journal of Physical Anthropology,* vol. 54 (1981). The quote on groups expanding via central Europe is on p. 138 of Bräuer's contribution to the new volume, *The Human Revolution: Behavioural and Biological Perspectives on the Origins of Modern Humans,* edited by P. Mellars and C. B. Stringer (Edinburgh University Press/Chicago University Press, 1989, in press). Bräuer's other paper was in *Journal of Human Evolution,* 10 (1981), p. 473.

Smith's comment on African features (or lack thereof) is on p. 195 of Spencer and Smith, eds., *The Origins of Modern Humans.* On interstratification, see *Nature,* September 25, 1980, vol. 287, p. 272. Trinkaus, who believes there was continuity in Asia too, said that admixture was demonstrated by the fact that modern humans, especially in the Old World, have very clear patterns of geographic variability, things like skin pigmentation, and he wondered how something like skin pigmentation, which he said had a low level of selection acting upon it, is so similar in arctic and South American Indians who have been around for 12,000 years or so with far less variability than in the Old World over a similar geographical tracking, proportionate to time. "The only way that we could get that is to have some significant degree of admixture between early modern and a local archaic population," Trinkaus says.

At around the time of Eve's earliest estimated mitochondrial appearance there were perhaps 1 to 3 million humans roaming the Old World. When sea levels were similar to what they are today, these increasingly technological people, both "moderns" and archaics, had at least thirty square miles per person across three continents over which to spread themselves, when all it might take to feed a single person was two to ten square miles of decent hunting-and-gathering land. If herds suddenly moved out of reach, or if disease and drought struck, or if, in the Asian areas, an isolated band of twenty or fifty humans found themselves against the wrath of a typhoon, it is easy to envision the elimination of an entire clan or band of clans. Karl Butzer believed that "many thousands of Acheulian isolates disappeared without issue in marginal environments or in the face of natural events beyond their capacity to cope." But entire populations of archaics that may have been entrenched for half a million years?

CHAPTER 18

A Holloway observation on Neandertal brain size is in *Ancestors,* p. 323. For pushing the date back to 9.2 million for the chimp-human split, see Spuhler's "Evolution of Mitochondrial DNA in Human and Other Organisms," in press, available through Los Alamos National Laboratory; his review article "Evolution of Mitochondrial DNA in Monkeys, Apes, and Humans," in the *Yearbook of Physical Anthropology,* 31:15–48 (1988); and his pamphlet published by the University of Queensland as its 1984 Butler Memorial Lecture, "The Evolution of Apes and Humans: Genes, Molecules, Chromosomes, Anatomy, and Behaviour." Wolpoff's quote on random survivorship is from p. 6 of his contribution to the upcoming Bräuer and Smith volume, tentatively *Continuity or Complete Replacement?* (in press).

The quote on fruitflies is from A. Latorre, A. Moya, and F. J. Ayala in the *Proceedings of the National Academy of Sciences USA,* 83:8649–8653 (1986). Wilson's intolerance to the idea of a slowdown is from p. 110 of Lewin's *Bones of Contention,* previously cited. In answer to a query on the gel, Wilson referred me to pp. 5945–5955 of *Nucleic Acids Research,* vol. 16, no. 13 (1988). The headline on a slowdown is from *Nature,* vol. 326, March 5, 1987, p. 93. See also Brown's study of urchins in *Science,* vol. 234, October 10, 1986, pp. 194–195.

The Gillespie quote is from p. 8009 of the *Proceedings of the National Academy of Sciences USA,* vol. 81 (December 1984). Britten's work, meanwhile, can be sampled in "Rates of DNA Sequence Evolution Differ Between Taxonomic Groups," *Science,* vol. 231, March 21, 1986. The important 1989 paper analyzing all the lab results and disputing Berkeley is *American Journal of Human Genetics,* 44:73–85 (1989). See also "Rates of Molecular Evolution: The Hominoid Slowdown," *BioEssays,* vol. 3, no. 1; "Stochastic Errors in DNA Evolution and Molecular Phylogeny," by Nei in *Evolutionary Perspectives and the New Genetics* (New York: Alan R. Liss, 1986), pp. 133–147; *Evolution,* vol. 37, no. 2 (March 1983), an article in this journal by Alan R. Templeton; *Journal of Molecular Evolution,* 25:330–342 (1987); Goodman, et al.'s, "Molecular Phylogeny of the Family of Apes and Humans (in press for *Genome*); and *Science News,* vol. 131, p. 74.

As for chimp-human divergence, three researchers from Japan, using mitochondrial DNA, had come up with yet another date, this one more recent: 2.7 million for the chimp split. What about *Australopithecus?* They hypothesized that perhaps there was interbreeding between protohumans and protochimpanzees, which would have allowed for the transfer of mitochondrial DNA after the protohuman had developed bipedality. Other Wilson papers: "Restriction Mapping in the Molecular Systematics of Mammals: A Retrospective Salute," Nobel Symposium, Karlskoga, Sweden, November 1, 1988 (in press); "Molecular Time Scale for Evolution," *TIG,* vol. 3, no. 9 (September 1987). See too "Origin of *Homo sapiens:* The Genetic Evidence" in *Major Topics in Primate and Human Evolution* (Cambridge: Cambridge University Press, 1986), in which the author argues that mitochondrial DNA does not accord well with nuclear evidence. ("We are left with the uncomfortable possibility that the new information on molecular anatomy may mean that we still know rather little about the genetic changes involved in the origin of man.")

CHAPTER 19

See "Maternal Inheritance of Human Mitochondrial DNA," by Wallace, Richard E. Giles, Hugues Blanc, and Howard M. Cann, in *Proceedings of the National Academy of Sciences USA,* vol. 77, no. 11 (November 1980), pp. 6715–6719, for some of Wallace's earlier work. For disease originating in mtDNA, see *Trends in Genetics,* vol. 5, no. 1 (January 1988), or *Cell,* vol. 55, November 18, 1988, pp. 601–610, or *Science,* vol. 242, p. 1427. Cavalli-Sforza's 1963 paper, co-authored with A. W. F. Edwards, was "Analysis of Human Evolution," in *Genetics Today.* That analysis of 235 people: *Proceedings of the National Academy of Sciences USA,* vol. 78, no. 9 (September 1981), pp. 5768–5772 (Wallace, Cavalli-Sforza, M. Denaro, H. Blanc, K. H. Chen, and E. Wilmsen). See also "Radiation of Human Mitochondrial DNA Types Analyzed by Restriction Endonuclease Cleavage Patterns," *Journal of Molecular Evolution,* 19:255–271 (1983). About the Chinese: see again Spuhler's piece in the *Yearbook of Physical Anthropology,* vol. 31 (1988), where he cites Yu, et al.

Nei's data and interpretations can be seen in Tomoko Ohta and Kenichi Aoki, eds., *Population Genetics and Molecular Evolution* (Japan Scientific Societies Press; New York: Springer-Verlag, 1985). See also Nei's "Genetic Relationship and Evolution of Human Races," in Max Hecht and Bruce Wallace, eds., *Evolutionary Biology* (Plenum Publishing, 1982). For linguistic affinities and genetic trends, see "Reconstruction of

Human Evolution: Bringing Together Genetic, Archaeological, and Linguistic Data," *Proceedings of the National Academy of Sciences USA*, vol. 85 (August 1988), pp. 6002–6006. In it Cavalli-Sforza had asked, "Which stimuli determined, and which technologies helped, expansions of modern humans to the whole Earth? It seems very likely that an important role was played by a biological advantage that may have developed slowly over millions of years and undergone a final step only with the appearance of modern humans: a fully developed language." Cavalli-Sforza had also said that the latest step in language "may have been an important factor determining the rapid expansion that followed the appearance of modern humans and the demise of Neandertals," which is what Wilson then echoed.

Nei rebukes those who found "Eve" in mtDNA on pp. 10, then 12, of Nei and Gregory Livshits's "Evolutionary Relationships of Europeans, Asians, and Africans at the Molecular Level," presented at a symposium in Tokyo in 1988 in conjunction with the award of the International Prize for Biology. Pope's writing was in the *Annual Review of Anthropology*, vol. 17 (1988), p. 62. Smith's rebuke of a localized speciation event was shared with Erik Trinkaus, et al., in "Modern Human Origins in Central Europe: A Case of Continuity," in press, *Aux Origines de la Diversité Humaine,* Nouvelle Encyclopédie des Sciences et des Techniques (Fondation Diderot) (Paris: Presses Universitaires de France). Klein's comment on archaics at Laetoli and Florisbad evolving from earlier populations from Broken Hill and Ndutu is on p. 35 of Klein's "The Stone Age Prehistory of Southern Africa" in the *Annual Review of Anthropology*, 12:25–48 (1983).

For a description of Laetoli Hominid 18, see *Nature*, vol. 284, March 6, 1980, pp. 55–56. Stringer's "It is only in Africa . . . " is from p. 294 of *Ancestors*. For Stringer's comment on northern Africa, see p. 120 of Spencer and Smith, eds., *The Origins of Modern Humans*. Bräuer's view on what happened during the Würm glaciation is on p. 158 of a paper he published in *Cour. Forsch. Inst. Senckenberg*, 69: 145–165 (Frankfurt-am-Main, 1984). I also rely upon Rightmire's "Africa and the Origins of Modern Humans," in *Variation, Culture and Evolution in African Populations* (Capetown: Witwatersrand University Press, 1986); and Stringer's lecture presented at a Chicago convention in 1987, as well as personal communications with all parties.

About Omo II and its archaic features: they put one somewhat in mind of old Rhodesian Man (broad and flattened frontal, long and low braincase, back of the vault strongly curved, with a large nuchal area suggestive of *erectus*), while Omo I had much more modern aspects, with a higher vault and more rounded parietals. Although Omo II is now thought to be an archaic *Homo sapiens*, Omo I is widely recognized as coming extremely close to the subspecies *Homo sapiens sapiens*. It was Yves Coppens of France who saw resemblances between Omo II and Solo men. And did Omo I, meanwhile, have a few vague resemblances to that archaic called Swanscombe? As for the Paris group's Y-chromosome work, see p. 408 of the *American Journal of Human Genetics*, vol. 38 (1986). One of the key researchers is Gérard Lucotte. The typing of 222 chromosomes is explained in a paper in press, "A Phylogeny of Human β-Globin Haplotypes and Its Implications for Recent Human Evolution," by Long, Aravinda Chakravarti, Corinne D. Boehm, Stylianos Antonarakis, and Haig H. Kazazian.

CHAPTER 20

In April 1989, Arensburg reported in *Nature*, vol. 338 (see also *The New York Times*, April 27, p. A28) that there was evidence Neandertal-like humans at Kebara could talk because one skeleton was found with the hyoid bone, a small U-shaped bone between the chin and larynx that anchors the muscles which power the larynx, lower jaw, and tongue. This meant to some that Neandertals and moderns may have spoken to each other if they met, since they had identical physiological equipment to do so. It also put another huge dent in Wilson's speech theory. On migration to America: a discovery of broken stone tools, bison bones, and charcoal at a farm in Oklahoma reported in 1989 implied human activity in the United States 26,000 to 40,000 years ago.

That Qafzeh was older than thought is "Thermoluminescence dating of Mousterian 'Proto-Cro-Magnon' remains from Israel and the origin of modern man," in *Nature*, vol. 331, February 18, 1988, pp. 614–616. The age was thought to be 92,000 years but literally updated in the *Journal of Human Evolution*, 17:733–737 (1989). On Ubeidiya: Bar-Josef thinks an incisor might be *erectus* but is suspicious of the skull pieces, which were found on the surface and are light for fossilized bone. Meanwhile Wolpoff says the pubic area in Q. H. 9 indicates male and the dentition is in the range of Neandertal, though overall he calls Qafzeh the best "non-Neandertal" for this time period. Stringer's comment on Qafzeh is taken from a paper in press, to appear in Erik Trinkaus, ed., *Emergence of Modern Humans: Biocultural Adaptations in the Later Pleistocene* (Cambridge: Cambridge University Press). Stringer also graciously left me an unpublished paper entitled "Replacement, Continuity and the Origin of *Homo sapiens*" upon a recent visit to New York. As to the tools: a rarefied book, yes, but one without page numbers. Various industries tend to mix. See Jelinek in *Science*, vol. 216, June 25, 1982, pp. 1374–1375. I also rely much upon Dorothy A. E. Garrod, et al., *The Stone Age of Mount Carmel* (Oxford: Clarendon Press, 1937), and the comment on "spear-like weapons" comes from p. 291 of this classic monograph.

The quote on Border Cave and its relationship to modern populations is from Peter B. Beaumont, Hertha de Villiers, and John C. Vogel in the *South African Journal of Science*, vol. 74 (November 1978). The remark on cultural development at Klasies before similar development in Europe is from Ronald Singer and John Wymer's *The Middle Stone Age at Klasies River Mouth in South Africa* (Chicago: University of Chicago Press, 1982), a monograph as excellent as the one from Mount Carmel. The quote is p. 206, and so is much of the background. Other tools: ecailles, microgravers, and diminutive scrapers. For the establishment of permanent communities before agriculture, see *The New York Times*, December 20, 1988, p. C1.

For genetic relationships among peoples, see Nei's previously mentioned "Genetic Relationship and Evolution of Human Races." For the distinctiveness of Jews, see p. 208 of *Nature*, vol. 314. As for Nei and Livshits on Europeans, Asians, and Africans, see their paper for the Tokyo symposium cited in the notes for the previous chapter. For the remark on genetic similarities between all peoples, see *Science*, vol. 177, August 4, 1972, pp. 434–436, results admittedly old. Irhoud, by the way, looked very much like Qafzeh Number 6. Some other places to consider for various reasons: Teshik-Tash, Kiik-Koba, Akhshtyr', Antelias, Dzhruchula, Ksar'Akil, Shukbah, Masloukh, Ras el-Kelb, Rozhok, Shubbabiq, Azykh, and always Shanidar. See Glynn Isaac and Karl Butzer, *After the Australopithecines* (Mouton Publishers, 1975) for environment of Africa back in prehistoric times. For Cro-Magnon, see Tom Prideaux, *Cro-Magnon Man* (New York: Time-Life Books, 1973). I also used as references: Philip Rightmire's "Middle Stone Age Humans from Eastern and Southern Africa," in press, to appear in Stringer and Mellars's forthcoming reference volume; Bernard Vandermeersch's "Les Hommes Fossiles de Qafzeh" (Paris: Editions du Centre National de la Recherche Scientifique); *Nature*, vol. 330, November 12, 1987, p. 159, for Kebara date; Trinkaus's paper in *L'Homme de Néandertal*, vol. 3, *L'Anatomie* (Liège, 1988), pp. 11–29; *Paléorient*, vol. 1 (1973), pp. 151–183; Wallace, et al., in the *American Journal of Physical Anthropology*, 68:149–155 (1985);

Horai, et al., in *Human Genetics* (Berlin: Springer-Verlag, 1987); Vallois and Vandermeersch in the *Journal of Human Evolution,* 4:445–455 (1975); and Nei in the *American Journal of Human Genetics,* vol. 26, no. 4 (July 1974). See also Nei in *Evolutionary Perspectives and the New Genetics* (New York: Alan R. Liss, 1986). Every time I looked for order, all that was to be seen was more complexity.

INDEX